Crawford Hollingworth is a behavioural change expert and co-founder of The Behavioural Architects, a consultancy that works with some of the biggest companies and organisations around the world. He writes and lectures regularly on how behavioural science can empower behavioural change.

Cathy Tomlinson is a behavioural researcher with years of consumer research experience. She led the Brain Team on their journey of discovery and empowerment.

T0187336

HOW YOUR BRAIN IS WIRED

An Owner's Manual

Crawford Hollingworth and Cathy Tomlinson

unbound

First published in 2023

Unbound
Level 1, Devonshire House, One Mayfair Place, London W1J 8AJ
www.unbound.com

'Small Kindnesses' by Danusha Laméris, in James Crews (Ed.), *Healing the Divide: Poems of Kindness and Connection*, Green Writers Press, 2019, is quoted by kind permission of the publisher.

While every effort has been made to trace the owners of copyright material reproduced herein, the publisher would like to apologise for any omissions and will be pleased to incorporate missing acknowledgements in any further editions.

Text design by PDQ Digital Media Solutions Ltd

A CIP record for this book is available from the British Library

ISBN 978-1-80018-186-1 (paperback)
ISBN 978-1-80018-187-8 (ebook)

Printed in Great Britain by CPI Group (UK)

1 3 5 7 9 8 6 4 2

We would like to thank each member of the Brain Team for bringing the science alive with their insight and inspiration.

Acknowledging that you do not have complete free will, or complete conscious control, actually increases the amount of free will and control you truly have.

John A. Bargh, *Before You Know It: The Unconscious Reasons We Do What We Do*

A Brief Look at How Your Brain Is Wired

No one is immune.

Our brains are wired to make us behave in certain ways:

⊙ to stick to the same old behaviour and avoid change, even though we like to think of ourselves as experimental

⊙ to hang on to an opinion or view once it has been formed, even though we like to think of ourselves as open-minded

⊙ to normalise what was once new and exciting, so we often get bored with what we have and continually seek out the new

⊙ to be future-focused, questing for more and failing to take time to enjoy the now

⊙ to operate on autopilot, which makes time rush by faster

⊙ to focus on and pay attention to negative things, ignoring the good stuff that makes our lives better

⊙ to encourage us in the belief that we are nicer and work harder than other people, even though we think that we don't have an egocentric bone in our body

Sound familiar? Our brains are contradictory creatures. Imagine how much richer and more rewarding life might be if sometimes you were able to short-circuit one or two of these wirings, even if just for a moment.

Brain-Rewiring Questionnaire:
Are you still wondering if this book is for you?

Do you sometimes think life has become a little too routine?	Yes/No
Do you take a lot of things around you for granted that you used to enjoy more?	Yes/No
Do you find time seems to go faster these days?	Yes/No
Do you sometimes not remember much about the routine things you do each day (your journey to work, for instance), almost as if you are on autopilot?	Yes/No
Do you ever find yourself thinking more about what you are going to say next in a conversation rather than listening to the person you're talking to?	Yes/No
Do you tend to like to keep things the way they are?	Yes/No
Are you sometimes daunted, even paralysed, into inaction by the size of the tasks you have to do?	Yes/No
Do you find small changes in behaviour hard to achieve, or habits hard to make or break?	Yes/No

Did you answer 'Yes' to any of those questions? Then *How Your Brain Is Wired: An Owner's Manual* might be just what you need.

Contents

A scientific journey though the brain's wirings, bringing alive the built-in biases that subconsciously shape your thinking and behaviour.

We lift the cloak of invisibility to reveal how they are subtly influencing our behaviour 24/7.

We examine how other people have been affected by living consciously with some of the brain wirings and offer lots of ideas for taking control. Discover how even the smallest changes can have a big behavioural impact.

Section 4 253

Putting your newfound awareness into action and taking control

Once you can begin to act more consciously on the awareness, to think about how to play to a particular piece of brain wiring or to try to block or counter another one, or to use a particular brain strategy to achieve a goal, however small, you will quickly see how small steps can lead to big changes.

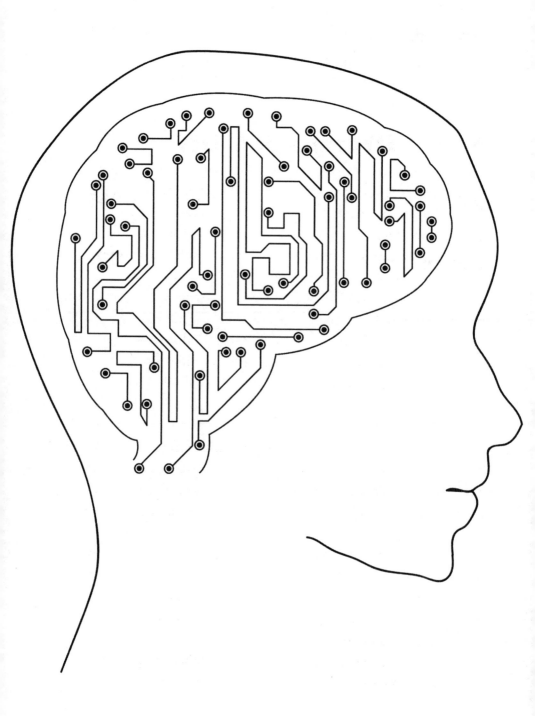

Introduction

In the last twenty years or so there have been exciting breakthroughs in our understanding of the human brain, many of which have revealed the degree to which our brains are wired subconsciously to direct or nudge our behaviour. The wirings influence how we think and prompt us to do the things we do. In other words, we're simply not as in charge of our thinking as we think we are.

Many of the brain wirings we discuss in this book are known as cognitive biases. These are dispositions or predispositions that cause us to think, behave and even remember in certain, often irrational, ways and are what ultimately make us fallible human beings. Some of these biases work on our sense of self and our place in the world. They can give us an inflated sense of our own importance, as the egocentric bias and self-serving bias* do, convincing us that we are the most important creatures in our universe, that our opinions are the correct opinions and that how we live and act is the right way to live and act. They can be useful in terms of survival and progress, but dangerous if allowed to flourish unacknowledged. There's confirmation bias too, which drives us to seek out only the information that confirms our own opinions rather than looking at the whole picture – a risky bias to defer to in a world chock-full of disinformation and fake news.

Some biases affect the way we remember past events; retrospective impact bias, for example, allows us to misremember

---

* These two biases can be treated as interchangeable.

past events in a more positive light, prompting us to repeat previous life choices and then make the same mistakes all over again.

And there are more – many more – biases, which behavioural psychologists have explored, tested and retested, defined and brought to life over the last few years. In fact, Wikipedia lists 185 (and counting). We'll look at fifty of these in the appendix.

The simple fact is that unless you are aware of, and understand, the way your brain is wired, every one of these biases will be impacting on you, subconsciously (that is, without your permission), all the time! Some will change the way you remember things, some will impact on how you make choices, and *all* of them will impact on how you behave. Until relatively recently, models of human behaviour and decision-making – generated without this understanding – were essentially based on the simple belief that humans were rational creatures who would make the right decision if presented with the right information. Here's what US psychologists Dan Ariely – who's well-known for his involvement in behavioural science – and Janet Schwartz write about this very subject:

> We often believe ourselves to be reasonable and rational beings. So we just have to have the right information to make good decisions, and we will immediately make the right decisions. We eat too much? Just provide calorie information, and all will be well. We don't save enough? Just start using a retirement calculator and watch our savings grow. Texting while driving? Just tell everyone how dangerous it is. Kids drop out of school. Doctors don't wash their hands before checking their patients. Let's just explain to the kids why they should stay in school and tell the doctors why they should wash their hands. Sadly, life is not that simple.[1]

It has been estimated that 95 per cent of our brain activity occurs at a subconscious level. And what's known as behavioural science has delivered a whole world of new scientific discoveries that reveal how our brains are wired – allowing us to understand, celebrate and even play to or leverage these wirings. It is these wirings that make us human beings, with all our strengths and weaknesses, but the next step in this incredible evolution of understanding and self-awareness is to accept the knowledge and use it to our advantage.

This sets an important context for this book because the more we know about these biases, the better equipped we will be to apply them consciously or to counter their subconscious influence. In fact, behavioural psychologists believe that just being conscious of their impact is powerful in itself. Awareness alone can empower us and help us to pay attention when maybe we are not being as rational as we thought or when we might not be making the best decisions, or to understand why we might find it such a struggle to make or break a habit, and then give us the tools for the job. One of the biggest insights, though, is that the more you understand the various brain wirings or biases and the science behind them, the more you will find yourself consciously trying to use them to your advantage. When conscious of the power of a wiring or bias, you can play to it, dialling it up or down, exploiting its power to make your life better in any number of ways or to achieve things you might have found hard in the past. Harnessing the power of the wiring or bias can be something of a light-bulb moment. The aim of this book is to keep that light bulb shining.

It will also come as no surprise that the power of these biases hasn't escaped the attention of governments, brands and companies who, as we will see in Section 2, are using these insights to nudge and steer our behaviour in every conceivable way. That's not to say that the nudging and steering don't have a moral and ethical intention at their core – helping us to make better decisions in all aspects of our lives – but morality and ethics are not necessarily *always* the prime motivators. Companies want us to hurry up and buy stuff. Governments want us to toe the line. It does us no harm to be aware of this; in fact, forewarned is forearmed. Take, for example, the crazy world of twenty-first-century politics: in this space we need our wits about us because political campaigns and campaigners are using every behavioural trick in the box to get us on their side.

The main aim of this book is to empower you by enabling you to understand how your brain is wired; to help you improve your life, make better decisions, achieve new goals; to understand yourself and your behaviour better. We will do this by making you aware of key subconscious wirings, biases and behavioural dispositions, bringing them alive in a simple, meaningful way, underpinned by some cool science.

You'll find numerous professors of psychology and the behavioural sciences quoted in these pages, together with references to published psychological studies, and they offer an excellent scientific basis for this rewiring journey. But interspersed with the academic voices are the voices and experiences of a group of individuals who took part in our brain rewiring experiment and road-tested how an increased consciousness of some of these biases played out in real life – our Brain Team. We recruited people who would challenge and inspire us (including some sceptics), and our Brain Team included administrators, artists, parents at home, mediators, lawyers, life coaches, solicitors, doctors, teachers, translators, software engineers, actors, a hypnotherapist, a drummer and a concert pianist! The aim was to see if they would derive benefits from this increased consciousness and the extent to which it would deliver empowerment and effect change in their everyday lives.

Their personal experiences along their journeys, from understanding brain wiring to the actions taken in response, are captured in Section 3. You will be left in no doubt that this reinvigorated brain awareness and understanding made a difference to their lives. Their experiences and reflections will also inspire ideas, actions and all manner of small ways to nudge and steer your behaviour.

*

This new understanding will empower you to:
⊙ be a better listener
⊙ be more conscious of your behaviour
⊙ check negative character traits
⊙ encode outgoing communications more effectively
⊙ decode incoming communications less pessimistically
⊙ approach situations with a more positive mindset
⊙ manage conflict more successfully
⊙ breathe new life into old
⊙ stop negative waves and trigger positive ones
⊙ have more fun

And here's a taste of the impact of this new understanding, as expressed by some of our Brain Team after their experiences as part of the experiment:

> *Everything you have planted in our minds over the past weeks has made me feel better, stronger, more in-touch, kinder, calmer, more lucid, a huge awareness of all around me. I embrace the insights. They are remarkably sustaining.*

> *It has opened my eyes to seeing and doing things in a more positive and productive way. You can learn a lot about yourself and others around you and find solutions to improve your life.*

> *I am enriched by this mindfulness and reflection. It has slowed me down, invigorated me, given me talking points for comparison and disagreement with friends and family. It encourages me to do things for myself, to take some risks, to be self-reliant. It reminds me how to be a better person if only I have the energy to pause before acting or to reflect before judging others and myself.*

I feel I have gained a greater knowledge of my behaviour and a way to cope with more things by learning to listen, change habits and not to stick to routine. If someone else was to take part I would encourage them to do it – to learn more about themselves and, in turn, learn more about others.

It could be life-changing. It has given me simple ways to understand how tensions or conflict might arise and made me more aware of what and how I say things.

Some of the stories and experiences that grew from the experiment will illustrate the biases from time to time. They are good stories, full of clarity, and they show that our brains have a greater plasticity than we may realise. But there is more to the experiment: after it finished, the volunteers occasionally got back in touch to report exciting developments in their lives that they attributed to the experience of thinking about these biases and consciously rewiring their brains. One ran a half marathon for the first time, in Paris; one trained for and embarked on a charity trek in Cuba; one started a new business; another, a full-time job; one applied for and got a place on a university degree course; one got a part-time job in a theatre (*'I grabbed the bull by the horns and went for it.'*) Others noted that their relationships had continued to strengthen – with their partners, children, parents. For others, the primary gain was a greater sense of calm, a more conscious and deliberate kindness to others, resulting in deeper feelings of contentment and a new sense of possibility. All noted a reinvigoration of life itself. No small thing.

There are countless books to help you redress the balance of your life – books about yoga and healthy eating, books about running well, books about finding the child within, books about wild swimming and decluttering your spare room. This book won't necessarily help you to do any of those things any better than you already can. Rather, it's about rewiring and rebooting your attitudes to stuff; about re-seeing yourself and your choices; and it will reveal something rather magical: how tiny tweaks to your behaviour can be all it takes to deliver a big, sometimes thrilling, reboot to your life.

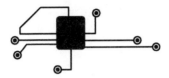

Section I: Let's start our journey by understanding how our brains are wired

In this section we will take you on a scientific journey around the brain's wiring, bringing alive the built-in biases that subconsciously shape your thinking and behaviour.

We are continuing to learn more from the behavioural sciences about why we think and behave in certain ways. From this learning, what has become crystal clear is that we are not cool, calm and rational thinkers who evaluate all information as it is presented before arriving at a carefully considered decision. (American psychologists sometimes call this mythical creature Chicago Man or Economic Man.) Constant cool and considered rationality is not our modus operandi. Many of us probably always suspected this; we just didn't know why. Now we do. What behavioural science has given us are simple frameworks and concepts that embrace and explain the irrational distortions that define human behaviour. These frameworks and concepts help us think more deeply about how and why we behave in one way rather than another. And this gets exciting, as it's something of a blueprint for human behaviour. Once you understand how something is constructed it's much easier to make changes.

The great thing about this new scientific understanding is that it makes simple, intuitive sense. You don't need to be a behavioural scientist to understand it. To give some order to the huge array of behavioural understanding and insight, we've defined four 'foundation stones' that express the key facts that behavioural science has taught us about how our brains operate. Once we understand these foundation stones, they can help us to think more deeply about how we behave and make decisions. And, as all our brains are wired in this way, it's a good starting point on the road to self-discovery, as well as giving an insight into other people.

Summary of the four foundation stones of behavioural science wisdom

1. **When we're thinking about something or making a decision, around 95 per cent of our brain is operating at a subconscious level.** When we see, hear or smell things, a mass of neural activity is triggered in our subconscious brain, and multiple connections and associations are engaged before we even have time to think about it!

2. **We use two systems for thinking and deciding.** System One is intuitive, spontaneous, emotional and fast. System Two is considered, thoughtful, effortful, slower. Does it come as a surprise to learn that we use our System One most of the time, even if we like to think we're using our System Two? And there's a good reason for this. System Two is hard to sustain; it's tiring and takes a lot out of us. In fact, we can only process about 40 bits of information a second in System Two, whereas System One can handle an impressive 11 million. In System One, we use lots of shortcuts to make quick, cognitively efficient decisions – what we might call gut reactions. If an instinctive feeling guides a decision, that's definitely System One at work.

3. **Context is ALL.** We hate making decisions in a vacuum and are always on the lookout for reference points to help us. We depend heavily on so-called anchors or sets of heuristics

(shortcuts or rules of thumb), which give our responses a head start and help us to plump for one thing rather than another. After all, we don't have all day. Our brains look for patterns, and every decision we make is made in a context: we search for anchors, familiarity, similar situations experienced in the past, things we know, what other people around us are doing – and then we throw them all into a metaphorical hat and make a choice. For better or for worse, a decision is made. Different contexts can lead to different outcomes, so a decision made in the heat of the moment is unlikely to stand up to scrutiny in the cold light of day. Ensuring you have the right context in which to ground decision-making or behavioural change goals is critical to their success.

4. **And then there are the cognitive biases.** These form the mass of subconscious biases prompting us to behave in certain, often seemingly irrational, ways. What other people do has a big impact on what *we* do – the power of the herd instinct is great; we love to stick together. We also love the status quo: it's so much simpler to keep things just the way they are.

These four foundation stones are deeply rooted in our lives pretty much all of the time. Whether we like it or not, they govern how we live, and for the most part they help us to make decisions quickly and, hopefully, efficiently.

So, on our journey into the wirings of your brain, let's look at each of the foundation stones one by one.

Foundation Stone One: Running on autopilot

A disconcerting 95 per cent of our brain activity occurs at a subconscious level; we're on autopilot and we can be easily influenced.

We do things without thinking about them all the time: breathe, walk, talk, drive, catch a ball – there's loads of stuff that we can handle on instinct. In the words of psychology and cognitive science professor John Bargh:

> Much of everyday life – thinking, feeling, and doing – is automatic in that it is driven by current features of the environment (i.e., people, objects, behaviours of others, settings, roles, norms, etc.) as mediated by conscious choice or reflection.[2]

Sheena Iyengar, Professor of Business at Columbia Business School, takes this idea further;

> Just like an iceberg, only a tenth of which is visible above water, our consciousness makes up only a small portion of our minds. In fact, the mind is more deeply submerged than the iceberg is; it's estimated that 95 per cent of mental behaviour is subconscious and automatic. Without conscious intervention, external forces can influence our choices with impunity.[3]

So, this means we're only living in a deliberative and conscious way a paltry *5 per cent* of the time.

The autopilot system is a wonderful thing that runs most of our behaviour, and apart from all the stuff it handles that's physiologically automatic – like breathing, digesting, healing, walking, etc. – there are all the other things we do without thinking too much, like getting dressed, making a cup of coffee and finding our way to work. Just run through a few things you do pretty much automatically. How many of us have taken the same journey home for years and sometimes arrive without remembering much about the trip? Autopilot is brilliant. We couldn't run our lives without it. But it services routine, regularity and sameness. It's why we often can't remember what we did last week, whether we shut the front door when we left the house or why the days blur into each other. If we forget to switch off autopilot and engage the *manual* controls at least some of the time, we won't be open to learning new things or perceiving things in a new way, and life might feel a little like *Groundhog Day*.

In his book *The Power of Habit*, former *New York Times* journalist Charles Duhigg describes a study that estimated that 45 per cent of the choices we make each day are based on habit (routine behaviours) rather than deliberate, thoughtful selection.[4] It sounds dull. If that is what we do, it probably makes *us* dull.

John Bargh's thesis was that 'most of a person's everyday life is determined not by their conscious intentions and deliberate choices but by mental processes that are put into motion by features of the environment and that operate outside of conscious awareness and guidance'.[5] It makes a person sound like a robot. But hang on a minute. Surely something can be done?

If we understand the power of autopilot and recognise when it is helpful for it to be active and when it will make dullards of us, it will help us to embrace it when we need it and allow us to escape from it every now and then. We'll be able to shift from automatic to manual, because we can learn to have control of the switch. One of the consequences of occasionally forcing yourself to be more consciously conscious is that it can make you feel as if time itself is slowing down, a wonderful and rare thing in a world rushing past us at hyper-speed. Have you ever noticed that when you are on holiday, or spending time in an unfamiliar place, it can seem like you've been there a week when it's only been couple of days? And that everything is new, you're paying attention, you're deliberately conscious? It works just like that.

One of the simple things we can do to redress our predisposition to autopilot is to spend some time just thinking about thinking. It helps us to be present in the moment, to make more active decisions and to remember more about what our lives are about. Thinking about thinking is also described as metacognition. It's when we pause to consider how we are thinking about something; how we are approaching a problem, for instance. When things are running to plan and life seems fine, we're unlikely to apply a metacognitive approach. But it's a useful technique: a momentary yoga-pause for the brain to take stock, to consider the moment. To think. And in so doing, to strive for the elusive 5 per cent of deliberative consciousness. A progressive US philosopher, educationalist and psychologist called John Dewey delivered a helpful explanation of metacognition in 1910 in his book *How We Think*. Dewey wrote:

> As long as our activity glides smoothly along from one thing to another ... there is no call for reflection. Difficulty or obstruction in the way of reaching a belief brings us, however, to a pause. In the suspense of uncertainty, we metaphorically climb a tree; we try to find some standpoint from which we may survey additional facts and, getting a more commanding view of the situation, decide how the facts stand related to one another.[6]

The power of subconscious priming

There's another angle to this subconscious living, and we can find that in 'priming'.

Priming is an incredibly powerful way to nudge or steer behaviour in a desired direction. It works at a subconscious level to subtly influence our behaviour, and it works on all our senses. Each one of our senses – sight, hearing, touch, smell and taste – can be triggered by an external stimulus setting off a mass of neural activity in our subconscious, tapping into our implicit associative memory.

Amanda Staveley is CEO of a financial advisory firm that negotiates billion-dollar deals with clients in the Gulf states – two male-dominated worlds right there. She describes the priming techniques she uses to build confidence:

> Before a meeting I'd listen to 'Eye of the Tiger' – it immediately puts you in a strong, confident mood. I also used to do the power pose thing, and I always ensured I was dressed professionally – a beautifully cut business suit can give a woman a lot of confidence.[7]

There's another perfect example of priming in the 2015 movie *Focus*, where Will Smith plays a con man who uses countless primes to trigger a gambler to choose a certain player on the football field. The player is wearing number 55 and the movie sequences a retrospective montage of all the ways the number 55 has been brought to the gambler's attention – without him knowing it: the doorman is wearing the number on his coat and people in the street have 55 on their shirts. It's a great example of the power of priming.

You can be primed by a given context to answer questions in a particular way; one might even say that your answers can be manipulated by a context. A good example of this is given by Richard H. Thaler and Cass R. Sunstein in their book *Nudge*, which aims to help the reader make better decisions about health, wealth and happiness.[8] College students were asked two questions: Q.1. How happy are you?; Q.2. How often are you dating? Asked in this order, the correlation between the two questions was quite low. But when Q.2 was asked first, it had a dramatic impact on responses to Q.1. Students who hadn't had a date recently suddenly became aware of how miserable they probably were.

Bringing alive the simple power of priming

Here are a few amazing examples of priming, in no particular order:

⊙ If you show people a smiley face on a screen – albeit so fast they don't even know they've seen it – they will like everything better that comes after.

⊙ In conversation, if you mention having been in a library, the people you are talking to will lower their voices.

⊙ If you expose people to money (a picture of money on a screen saver, for instance, or a pile of money on a desk) and then accidentally drop a big bundle of pencils all over the floor next to them, they will pick up fewer pencils (i.e. be less inclined to help you) than people who have not been exposed to (primed with) money. Another study that explored the priming effect of money revealed other angles on this hard-nosed behaviour. Apparently, thinking about money makes us:
  ◎ more selfish and self-reliant
  ◎ less likely to ask for help from others
  ◎ want to spend more time alone
  ◎ more likely to select individual tasks rather than those that require teamwork
  ◎ choose to sit further away from others

⊙ If you put the word 'VOMIT' on a screen, within a couple of seconds people will make an expression of disgust and physically recoil – but they won't necessarily know they have done this.

In his book *Thinking, Fast and Slow*,[9] Economics Nobel Laureate Daniel Kahneman describes a brilliant honesty-box experiment run by the psychology department at Newcastle University. It's a study that has been called into question by some, since its results can't always be replicated, but it remains a powerful example of the power of priming. In the experiment, there's a coffee room with prices for drinks and a payment box, but no one checks to make sure people pay the right money. Each week, the Newcastle University researchers posted a list of prices above the honesty box and, while the prices themselves stayed the same, each new list had a different picture at the top – sometimes it was a picture of flowers and sometimes it was a pair of eyes looking right at you. They tried out different kinds of eyes too. Results showed that people paid more for their coffee in the 'eyes' weeks than the 'flowers' weeks and that the scariest eyes pulled in the most money. The concept of being watched is a powerful prime for honesty.

It works on crime prevention too. Newcastle University also used eye posters to tackle bicycle crime and discovered thefts fell by 62 per cent over the two years the eye posters were in place. (Where there were no posters, thefts rose by 63 per cent.) According to the Newcastle team (who would seem to have fixed on eyes as their go-to security system), the idea was adapted from a previous experiment in a university canteen, where images of staring eyes were used to encourage diners to clear away their trays.[10]

The study's authors concluded that there is definitely something to be said for the presence of watching eyes as a deterrent for antisocial behaviour, especially in situations where there are only a few other people around. When there's less and less money to pay for actual police officers, perhaps posters are the answer?

In Nottingham city centre a similar poster campaign, with the words 'THIEVES: WE'RE WATCHING YOU' below a pair of fiercely staring eyes, drove home the message across the city. Robin Dunbar, evolutionary psychologist and anthropologist at Oxford University, cited this example in a lecture in March 2012. He said the campaign had resulted in a 17 per cent reduction in street crime.[*]

---

[*] Although the Freedom of Information section at Nottinghamshire Police were unable to confirm this figure, they did acknowledge that 'a poster was printed and distributed locally to try to prevent offending'.

You can be easily primed by other non-verbal cues. The authority bias primes via the cues delivered by uniforms associated with authority and trust. Police officers and doctors in uniform prime you to accept what they have to say more readily – and it's the anchor bias that helps us to fix on cues like uniforms and build our subsequent judgements upon them. (We have all heard stories reporting the actions of unscrupulous people who have appropriated official status and used it to prey on others, relying on the fact that they will be believed.) Clothing can also signal expertise and excellence; after an Under 16 girls' hockey tournament, a member of one of the teams spoke of the winners with awe: 'They had such amazing kit ... They all had numbers on their backs – they were like professionals. Our kit was rubbish.'[11]

Check out this excerpt from an article called '9 Reasons Why Your Looks Matter: Why Give A Damn About Appearance', taken from the Real Men Real Style website but the same holds true for women too:

1. Clothing is the primary instrument in creating a positive first impression

People are superficial, not just as a cultural phenomenon but as a hardwired instinct going all the way back to when our brains needed to make snap judgments on what was a stripy rock and what was a tiger about to eat us. We tend to be done formulating our initial opinion of someone before we've actually spoken to them ... That means that your clothing is going to have a direct effect on people's default assumptions of you – the better you're dressed, the more respect and attention they're going to automatically give you.

2. Clothing can increase your perceived status among your peers

Even after the first impression is over your clothing can help improve people's reactions to you. Society is very visually-based, and better-dressed men routinely experience better treatment and service than their sloppier counterparts. Clothing serves as a substitute for character in the eyes of people who don't know you well enough to judge you by anything else ... That may sound superficial, but it's true whether we think it should be or not. Your appearance may not mean much to *you*, but it does to the people who see you every day, making it worth caring about.[12]

Experiments have shown you can manipulate people to behave in a particular way by carefully priming them to do so. In fact, so many experiments have been carried out to prove this that it's hard to pick just a few to illustrate the concept. The significant element in all of them is that the participants were completely unaware that their behaviour had been managed in any way and would, in fact, adamantly deny that their behaviour had been influenced at all. Priming was carried out in such a way that they were simply not conscious of its effect.

Primed to be nasty or nice

While it's not entirely clear why anyone would want to prime someone for nastiness, the following experiment certainly demonstrates that nasty priming works. Participants were primed to behave rudely by being given sentence-unscrambling exercises associated with rudeness; the scrambled sentences they were given included words like 'interrupt', 'assertive', 'disturb'. At the same time, other participants were primed to behave more politely; their sentences incorporated language associated with politeness – 'respectful', 'patience', 'polite'. A further control group were not primed in any specific way.

Having completed the word exercise, participants were asked to go to another room, ostensibly to take part in an unrelated experiment. When they entered the room, they had to report to the person leading this experiment. However, in each case, the leader was deep in conversation and, for the purposes of the experiment, would remain so for a full ten minutes. The real point of the experiment was to see which of the participants would interrupt soonest. A laudable 84 per cent of the 'politely primed' participants waited the full ten minutes without speaking and 62 per cent of the control group were able to wait it out too. What of the rudely primed group? A mighty 66 per cent of them interrupted long before the ten minutes was up.[13]

And who knew that eating sweet things can actually make you sweeter? According to sensory and cognitive neuroscientist Rachel Herz, even a brief taste of sweet food can make people feel more agreeable and more likely to volunteer their time to help others. In fact, Herz goes as far as to say that a sweet tooth is an indicator that an individual is more likely to be cooperative, kinder and more compassionate.[14]

A hot drink can be such a comfort, but more than that, it can trigger generosity and an inclination to kindness – something to consider if you want to win someone round or are asking them for money. In an experiment run in the psychology department at Yale University, participants were met individually by a member of the study team to travel together in the lift for a few floors.[15] The team member was holding some books and a cup of coffee but needed to ask a few preliminary questions on the way up. *Would the participant mind holding the coffee cup?* Half the participants were given a hot cup of coffee to hold, the other half an iced coffee. When they reached their destination, they were asked to read a short description of a stranger and then to evaluate the stranger's likely personality based on what they had read. Would it surprise you to find that those who had been given a hot cup of coffee rated the stranger as more generous, more sociable, happier and better natured than participants who held the iced-coffee cup?

In a related experiment run by the same team, student volunteers were asked to assess a therapeutic hot or cold gel pack (the kind used for pain relief). After the evaluations, the volunteers were offered a reward for taking part and could choose either to keep the reward themselves or to give it to someone else. Those who had held the hot gel pack were more likely to choose to give the reward to a friend. Psychologist John Bargh, who co-authored the paper, concluded: 'It appears that the effect of physical temperature is not just on how we see others, it affects our own behaviour as well. Physical warmth can make us see others as warmer people, but also cause us to be warmer – more generous and trusting – as well.'*[16]

Primed to behave in an elderly or youthful way

The following example has been questioned by some psychologists, who have been unable to replicate the effects first found by John Bargh and colleagues. But it's interesting nonetheless and it has been shown to work in some instances. The concept is that you can manipulate people to respond physically in a particular way by priming them with stereotypical concepts of behaviour. In another experiment, some participants were primed with stereotypical concepts of the elderly when words like 'Florida', 'sentimental' and 'wrinkle' were included in a series of four- or five-word scrambled sentences that they were required to reassemble in an appropriate word-order. Having completed their word task, they left the room to walk to another room further down the corridor. The length of time it took them to complete this journey was measured surreptitiously – and results showed that the participants walked noticeably more slowly, more like old people, in fact, than others involved in the experiment who had *not* been primed in this way. Those who had been primed

---

*   There have been problems replicating this experiment with the same results, but it seems to make intuitive sense, don't you think?

with elderly stereotypes also remembered less about the features of the rooms they had been in.[17]

In Seattle, a visionary retirement community, home to more than 400 older adults, established itself as an Intergenerational Learning Center, effectively a preschool where older residents and preschool-age children are given the chance to bond. Children and older adults come together five days a week for a variety of activities including music, dancing, art, storytelling and socialising. Everyone benefits and the children are primed to learn about ageing, disabilities and diversity. As well as giving and receiving quantities of love and attention, the older adults are primed with a renewed zest for life – to feel more youthful, in fact.[18] A similar experiment was run in the UK, albeit briefly, mixing ten preschoolers with ten older adults (all aged over eighty) for six weeks. A comparison of medical tests on the older adults, covering mobility and attitudes to life, conducted before and after the six weeks, revealed improvements across the board, with heightened moods, more optimism and greatly enhanced mobility. A TV documentary of the experiment made for joyful and uplifting viewing.[19]

Primed to be successful

Cognitive neuroscientist Sara Bengtsson devised an experiment in which she tested students' performance in cognitive tasks (at the same time scanning their brains), having first primed them with either positive or negative performance expectations.[20] To induce expectations of smartness and success, she primed college students with words such as 'smart', 'intelligent' and 'clever' just before asking them to perform the test. To induce expectations of failure, she primed them with words like 'stupid' and 'ignorant'. The students performed better after being primed with the positive message and not so well when primed with the negative one. Examining the brain-imaging data, Bengtsson found that the students' brains responded differently to any mistakes they had made depending on whether they had been primed with expectations of success or failure. When a mistake was made by the success-primed participants, she observed enhanced activity in the anterior medial part of the prefrontal cortex (a region that is involved in self-reflection and recollection). However, among participants primed with failure, there was no heightened activity after a wrong answer. It appears that after being primed to be 'stupid', the brain expected to do poorly and therefore showed no signs of surprise or conflict when it made an error. Those whose brains were anticipating success became aware of any mistakes and worked to correct them; their brains effectively sent them a warning to tell them if an answer was wrong. Those who were not expecting to do well received no such signal. Perhaps schools could add positive 'smart priming' sessions to the curriculum in the lead-up to public examinations, or just as a general confidence booster.

Primed by nominative determinism

Whether nominative determinism is a great example of priming or not, it gets to be a part of this section ('The power of subconscious priming') because it's too prevalent to be simply a coincidence. Nominative determinism is where your name influences something about your life – the job you choose, the person you marry, how successful you turn out to be. The term was coined by *New Scientist* journalist John Hoyland in 1994 and it has been further explored by psychologists, who have noted that it's a form of 'implicit egotism' in that things that are connected to oneself trigger preferential associations.[21] We like ourselves and we like things that remind us of us. Take the list of humorously occupation-appropriate surnames compiled by a researcher at the University of Arizona. It celebrates the surnames of medical professionals and includes the splendid Drs Root, Pain, Fear, Rensch, Fang, Tusk, Eke and Toothaker (dentists); together with Dr Rash (dermatology), Dr Peek (optometrist) and Dr Strange (psychiatrist).[22]

Richard Wiseman, Professor of the Public Understanding of Psychology at the University of Hertfordshire, conducted a study that demonstrated that we are more than likely to make a judgement about someone before we meet them, based purely on their name.[23] In 'The Name Experiment', he asked 6,500 participants to determine which of the twenty male and twenty female names they anticipated would belong to the people who were the most successful, attractive and lucky. It's perhaps no surprise that traditional (and royal-sounding) names like Elizabeth, James, Caroline and Richard were expected to be most successful.

So, names trigger expectations. This has negative connotations where there is any likelihood of racism coming into play. A study conducted by David Figlio, Dean of the School of Education and Social Policy at Northwestern University in Illinois, demonstrated that African American children whose names easily identified them as such ran the risk of being judged as 'linguistically low status', more likely to be treated differently by teachers and less likely to be considered or expected to be gifted.[24] The same study revealed that Asian children with 'racially identifiable' names were, conversely, *more* likely to be treated as gifted since Asian names signal attributes associated with success. This is evidence of implicit racism and Figlio's conclusions suggest there is a role for training to increase teacher awareness of their apparent tendency to treat African American students differently based on their names. Increasingly, perhaps, the use of anonymised tests, CVs and applications would seem to be the only way to avoid this kind of (un)conscious bias.

Primed to clean up

And here's another area of powerful priming to consider: priming by scent. You know how supermarkets use the smell of baking bread not just to encourage you to buy bread but to infer all sorts of good things about the freshness and quality of the food products they offer and their whole retail ethos? And how estate agents tell us to fragrance our houses with the aroma of freshly brewed coffee if we want to sell them? Well, there's no doubt that olfactory priming works.

You might want to try this one at home – it seems pretty useful to me. Psychologists at the Dutch universities of Utrecht and Radboud asked students participating in their experiment to complete a questionnaire while sitting in a cubicle. Some students sat in a cubicle where there was a bucket of water fragranced with a lemon-scented all-purpose cleaner; other students sat in fragrance-free cubicles. Once the students had filled in the questionnaire, they were taken to another room for refreshments. The refreshments included a particularly crumbly biscuit.

It was this biscuit-eating moment, not the questionnaire that they had completed, that was the point of the experiment. The students were filmed eating the biscuits, the hypothesis being that those participants who had been exposed to the smell of a cleaning product would display a generally tidier approach to their snacking. And so it turned out to be. In comparison with the control group (who had not been exposed to the cleaning product), they cleaned up the crumbs significantly more often, the very smell of cleanness having primed them to behave in a clean and tidy manner.[25] Imagine the implications for teenagers and their bedrooms…

Primed to keep on keeping on

How about this for an easy but powerful application of priming? A simple smile is all it takes. In the face of physical exertion, smiling away the pain really works. British sport psychologists have found that smiling when the pain kicks in during a long-distance run can boost a runner's stamina by 2 per cent.[26] According to researchers, 'smiling could induce a relaxed state without people consciously trying to relax'. They found that 'intentional smiling' reduced 'effort perception'. Try it. Try it when things are getting you down, too. Change your expression from grim to grin and you'll cheer right up. (A word of warning, though: in contrast, frowning increases tension.)

Foundation Stone Two: System One and System Two thinking

This is a very useful dual-thinking model: you have the fast, automatic, intuitive, emotional System One, which reacts to cues and looks for patterns, and the slower, more effortful, logical, deliberate System Two, which explores possibilities and probabilities.

We'd all like to think of ourselves as the kind of people who think things through in a System Two way. We wouldn't make a snap decision about something serious; we'd weigh up all the pros and cons. That is, however, where we are wrong. We mainly go with our gut, we make emotional decisions, we base our judgements on past experiences, we see what's right in front of us and fail to look beyond that. We're all fully paid-up members of System One, especially when we're short of time or energy and when cognitive ease is what we seek. And who can blame us? It's tiring to use System Two, and no one could do it all the time. System Two is slow, it wants to explore all the options, to deliberate; it finds it hard to make decisions, and we haven't got all day. Plus, System One is much harder-working – if you were to think about the systems in terms of how much information they can process, it's 11 million bits per second for System One versus 40 for System Two. System One wins hands down.

So, what to do with this information? Here are some helpful tips.

System One is a good thing: it builds motivation and meaning. It protects us from danger, allows us to respond quickly and intuitively, and gets on with the job; it lets us make decisions fast, based on past experiences and learning. System One is behind our unconscious tendency to imitate the posture and behaviour of people with whom we interact (even that of complete strangers). Research has found that physical mimicry of this kind facilitates smooth social interactions and increases liking – and we don't even know we are doing it!* (Imagine! You can get someone to like you just by imitating them!) In a study designed to test this, participants were invited to take part in an experiment that was ostensibly based on looking at photographs.[27] Each participant sat at right angles with a confederate (someone who knew what the experiment was *really* setting out to test), and looked down at a number of photographs (minimising their ability to make eye contact). In some of the sessions, confederates rubbed their faces repeatedly, while in other sessions they shook their feet (and, of course, there were control sessions where they kept their hands and feet still).

Video playback of the one-on-one sessions revealed that participants who were teamed with the face-rubbing/feet-shaking confederates respectively rubbed their face or shook their feet significantly more often than those in the control sessions. The final aspect of the test required participants to rate the 'likeability' of the confederate with whom they had worked. In all cases where face-rubbing and foot-shaking had been initiated and, crucially, mimicked, confederates were rated as significantly more likeable than in the partnerships where actions had not been emulated. At the end of the experiment, when participants were asked about the face rubbing and foot shaking they had engaged in, none had any awareness of doing so. 'In a chameleon-like way, the participants' behaviour automatically changed as a function of their social environment,' as the researchers put it in their findings.[28]

If you want to make a good impression on someone or get

---

* A strong case for the old maxim 'Imitation is the sincerest form of flattery.'

31

someone to like you, try subtly copying their gestures. (It's amazing how often you find yourself doing this anyway, once you're aware of its power: I was chatting to a friend in the street recently and realised that we were both nodding, smiling and ruminatively rubbing our chins!)

But System One doesn't allow for stuff beneath the surface; it takes things at face value instead, so there's a superficiality at heart that we should be aware of, and it's less open to *new* thinking, new ways of doing and being, and likely to be party to helping us remake old mistakes. Each time we pay off the minimum suggested balance on our credit card, we're in thrall to System One.

It's prey to all kinds of biases, too; it's System One, for instance, that allows us to hang on the coat-tails of confirmation bias when we're researching an argument, finding all the data that supports our views and disregarding the stuff that doesn't.

There are simple studies that show how the cognitive ease of System One can actually shape our thinking. One Dutch study required one group of students to recall three instances of using their bikes, while another group was asked to list eight. Afterwards all the students were asked how often they used their bikes. Those who had been asked to recall three instances reported using their bikes more often than the others. This was because it was harder for the 'eight' group to remember eight instances, so they deduced they didn't use their bikes that much, whereas the 'three' group found it easy to think of three instances and concluded that they used their bikes pretty often. Both conclusions were influenced by how easy it was to answer the initial question.

You'd imagine judges would be the most deliberate and constant users of System Two thinking because other people's lives and futures are in their hands. They're not. Study after study has been conducted with judges (ranging from the inexperienced to those with years of judging under their belts), to determine their freedom from bias. In one study, prior to passing sentence on a defendant, judges were given certain additional information – either a plaintiff's request for compensation (where some judges were asked for high amounts and others lower amounts) or random demands for the length of a custodial sentence (higher or lower) from the prosecution, with both the compensation and custodial amounts having simply been determined by the roll of a dice.[29] The information was usually given to the judges purely randomly, and they were told as much – in other words, they knew they were taking part in an experiment. The aim of the study was to see to what extent the judges anchored on the information. In every case, the high or low figure the judges were given (either as a compensation amount or length of custodial sentence) impacted on their sentencing. They gave more compensation or longer custodial sentences where the figure was higher, and less compensation and shorter custodial sentences where it was lower. Their System One thinking prompted them to anchor on a suggestion. It was the easiest thing to do.

If you want to manipulate someone, or at least get them around to your way of thinking, make sure you make things easy to follow, easy to understand, easy to respond to, keep them going with the flow. Make things difficult and, as John Dewey pointed out, we'll be forced to pause, to (metaphorically) 'climb a tree', and, in so doing, we might bring System Two to bear on the issue. Social media postings that lead with images are playing to the cognitive ease of System One thinking – we haven't got time to click on a link to a discursive article, but we have got time to click on the image or to watch an image that plays automatically on our screens without any interference from us at all. (It's one of the reasons why Instagram is such a hit.)

There's another worrying side to the ease of System One and it's how it impacts on our political thinking. Did you know that the less thoughtfully we consider things, the more likely we are to lean towards the political right? Thinking with our gut makes us politically conservative. A study conducted by psychologists in the US examined a hypothesis that low-effort thought leads to political conservatism.[30] They found right-wing opinions intensified when participants failed to give their wholehearted deliberation to the thinking process; when they were drunk, for example, when they were distracted, when they were under time-pressure or when they were instructed not to think too hard. Conversely, when participants were asked to deliberate more deeply, they shifted to the left.

Politicians who use short sentences, short words and chants to incite the crowd are playing to cognitive ease. Politicians who talk articulately about the economy require concentration from their audience. Donald Trump's repeated use of 'Drain the swamp' and 'Build the wall' reached the electorate in ways that Hillary Rodham Clinton's eloquence on the economy did not. And witness Trump's supporters' enthusiastic uptake of the vicious monosyllabic chant 'Lock her up!'

In December 2017, a *New Scientist* leader article called time on effortless thinking, describing it as being 'at the root of the world's most serious problems: xenophobia, terrorism, hatred, inequality, defence of injustice, religious fanaticism, and our susceptibility to fake news and conspiracy theories'.[31] It went on to say that all of those problems are 'facilitated by people disengaging their critical faculties and going with their gut – and being encouraged to do so by populist politicians channelling anger at the liberal establishment'.

A quick sidebar on the problems of drunk thinking. It's not just the potentially dangerous shift to a right-wing ideology or two that you've got to look out for. There's the drunk dialling/texting thing that seems so tempting at the time. But help is at hand. Turns out there are apps to help us deal with our stupid drunk-thinking decisions – apps like DrunkBlocker and Drunk Mode, which allow you to block certain numbers on your phone, e.g.

ex-boy/girlfriends who you're likely to end up calling thirty-seven times after one too many drinks on a Saturday night. The Drunk Mode app can be disabled if you can solve a basic maths problem, or you can nominate a time period for blocking, making the data in your phone's address book inaccessible until you relaunch the app after the allotted time is up.

We need to recognise when to deploy System Two thinking;[*] when we need to use our effort and strength to think hard. We never actively deploy System One, but we must be able to override it. That said, keep in mind that System One and System Two are two modes of thinking that operate together, not in isolation; usually this is cognitively very efficient, but not always!

Foundation Stone Three: Context is ALL

Behavioural science has spent a long time looking at how the context in which information appears can radically influence behaviour. In this book we are going to treat this concept loosely and explore it from three different angles. We will look at how intentions made in one state or context may not carry through to another state or context, and how this means good intentions made in the cold light of day might get lost in the hot reality of the moment. We will also look at how, in any decision-making process, we seek anchors or reference points to guide us and may rely on these too heavily. We will also examine the powerful concepts of *framing* and *chunking* and how the different way information is presented can have a radically different impact on behaviour.

---

[*]   One highly recommended way of doing this is to read Daniel Kahneman's book *Thinking, Fast and Slow* (Farrar, Strauss and Giroux, New York, 2011).

The hot–cold empathy gap OR *Where good intentions can go to die*

A cold state is where we are calm, collected and rational. A hot state is where we are emotional, excited, hungry, in pain or aroused in some way. Research has shown that humans have a 'hot–cold' empathy gap – in other words, when we are in a 'cold' or rational state we can be poor predictors of how we might behave when we are in a 'hot' or emotional state, and vice versa.

It's why we can easily vow never to drink again in the ('cold' zone) morning, when we feel a little fragile after a ('hot' zone) big night out. And yet, once we've recovered and we find ourselves in a hot zone all over again, we conveniently forget all about our plans and promises. It's why 'in the heat of the moment' is an expression that conjures all sorts of actions and possibilities that would be unimaginable 'in the cold light of day'.

Never go shopping for food when you are hungry (a 'hot' state) – it's wise advice and based on scientific research that shows that the hungrier we are, the more likely we are to choose less healthily and seek foods with more calories. In fact, John Bargh would tell you never to shop for *anything* when you are hungry as it will encourage you to buy more of *everything*, not just food.[32]

In a study among thirty-five male college students at the University of California, Berkeley, researchers found that opinions on sexual behaviour varied considerably depending on whether an individual was in 'the heat of the moment' or not.[33] Test results concerning the sexual preferences and appetites of the students varied hugely depending on whether they were in a 'cold, rational, unaroused' state or in a more aroused, sexual state (don't ask). When questioned about their sexual preferences in a cold state, the majority said they would always use a condom and were not interested in a threesome, having underage sex, spanking or tying their partners up, and so on. But, in an aroused frame of mind, the answers the men gave were considerably less constrained. For example, when asked in a non-aroused state if they would like to tie up their partners, 'only' 47 per cent answered yes. When aroused, 75 per cent said they would like to do this. The authors of the study concluded that expecting willpower alone to 'protect' us from our heat-of-the moment predilections would probably be unwise, and that we should instead adopt a more practical mindset: avoiding situations where we would find it hard to resist something. So, don't go to the party if you don't want to drink or smoke, for example, or have back-up protection (take a condom) should willpower alone not be enough.

We are mercurial in our natures – ask us at 7 a.m. to make a lunch choice and we'll make a healthier selection that we do at noon; at 7 a.m. we're not tired and frazzled from a morning's commute and the demands of our working day. At noon we are exactly that. Thomas Schelling, a renowned US economist who won the Nobel Prize in Economic Sciences in 2005, described the dichotomy of states thus in his 1980 essay 'The Intimate Contest for Self-Command':

> People behave sometimes as if they had two selves, one who wants clean lungs and long life and another who adores tobacco, or one who wants a lean body and another who wants dessert, or one who yearns to improve himself by reading *The Public Interest* and another who would rather watch an old movie on television. The two are in continual contest for control.[34]

Research by George Loewenstein addressed the implications of the hot–cold empathy gap on medical decision-making.[35] He showed how the fact that we are so often incapable of imagining the way 'ill' feels when we are perfectly well, and, conversely, how 'well' feels when we are ill, impacts our ability to take medication. This is particularly significant in the case of those with bipolar disorder, high blood pressure or diabetes – conditions for which medication must be continued even if no obvious symptoms are apparent (diabetes is known as a silent killer for this very reason). We understand the concept of taking medication when we have tangible symptoms (the hot zone, if you like), but it is a more difficult protocol to sustain when we have no noticeable symptoms to treat.

The effect of the hot–cold empathy gap on medical practitioners is perhaps the most unsettling. Pain-free doctors who must prescribe pain-relief medication for suffering patients may well either *under*prescribe because, as Loewenstein puts it, they 'underappreciate their patients' pain' or, indeed, *over*prescribe because they 'over' appreciate it. And, in the case of end-of-life care, the hot–cold empathy gap prompts other concerns. In a study by consultant oncologist Maurice Slevin and colleagues, (healthy) oncologists had to make a judgement on how they would treat *themselves* if they were end-stage cancer sufferers having to weigh up the unpleasantness of a course of chemotherapy against the promise of more life.[36] Most of the oncologists judged an earlier death a better option for themselves than the course of chemo in that hypothetical situation (a judgement they would presumably consider appropriate for their patients too), with only 6 per cent of oncologists opting for the chemo. In contrast, two fifths of actual cancer patients – those who really were facing the prospect of less life – opted for chemo, regardless of its heavy toll upon them.

So, if we make a decision in the cold light of day or, indeed, in the heat of the moment, we should at the same time think about how we can create structures around us, strong supportive contexts, that will enable us to sustain that decision when the temperature fluctuates – think empathy-gap closure.

Anchoring: We're always looking for footholds

When making a decision, people look for anchors or reference points they know and can rely on, and work from there.

Anchoring describes the common human tendency, when making decisions, to seek out, and often rely heavily on, one trait or piece of information. Your brain looks for anchors – intense and/or distinctive pieces of information it can fix on – and judges everything else from that point. So, changing the anchor or reference point can radically change your opinion or how you decide to act.

In a restaurant, we might anchor on the price of the cheapest and most expensive bottles of wine on the wine list and then work up from the cheapest and down from the most expensive, adjusting our level between the two extremes and often then selecting a bottle that is one or two above the cheapest. This is sometimes called extremeness aversion.

Here's a good example of how extremeness aversion/anchoring works; it's taken from a study conducted by cognitive psychologist Amos Tversky and marketing guru Itamar Simonson.[37] In the study, one group was asked to choose between two Minolta cameras, one costing $170 and the other, better camera, costing $240. People's choices were split 50/50 between the two options. A second group was given three cameras to choose from: the $170 version, the $240 version and, in addition, a high-quality Minolta camera costing $470. How did these participants decide? Well, the impact of extremeness aversion results in a much greater popularity of the $240 camera as people select away from each of the extremes – 57 per cent chose the $240 version with the rest split equally between the top and bottom ends of the scale. It's fascinating to see how our brains can recalibrate and recalculate desirability so fluidly.

Sometimes our perceptions of reality prompt us to anchor rather shakily on what turn out to be only perceptions, and we should beware of this. A study run by Daniel Kahneman and David Schkade, who specialises in the psychology of judgement and decision-making, asked a sample of Californian and Midwestern students how much happier they believed Californians to be compared to Midwesterners.[38] All the students said they believed Californians must be considerably happier. However, there was no difference between the happiness rating of actual Californians and Midwesterners. The study revealed that the bias lay in the fact that most of the students focused on, overweighed and overestimated the sunny weather and ostensibly easy-going lifestyle of California and devalued and underrated other aspects of life and determinants of happiness in the Midwest, such as low crime rates and safety from earthquakes. It's something we do when we imagine the lives of our peers: we anchor on the perception that their lives are almost certainly better and more fulfilling than ours; they have more friends and spend their weekends flitting from one fabulous social engagement to the next, don't they? The reality is just California dreaming, all over again.

When it comes to gauging a context for information and building up a bigger picture, anchoring can help to deliver facts in a useful way. The murder-rate per head of population in El Salvador was 61.7 per 100,000 in 2017. How do you judge whether this was a normal level of murder? Anchor it against the murder-rate per head of population globally in 2017: 6.1 per 100,000.[39] Job done.

Framing: Change the frame, change the picture

The way in which information is presented, received or interpreted by us can completely change how it is viewed and the effect it creates. Providing the same information in a different way, or turning something on its head to see if there's another interpretation to be had, can change how we think about something and can make a tricky task more achievable. This is called framing and reframing and it's an incredibly powerful tool that can help in many aspects of everyday life.

If you actively try to think differently about something, you can bring about a change in an experience. We are more in control of how we conceptualise an experience than we know ourselves to be. In many situations we have the power to make a bad thing better, a hard thing more achievable, or what can be seen as dead or wasted time as time better spent!

Take a look at this situation that originally appeared on the website of the Royal College of Psychiatrists, and see how a *helpful* reframing of the *unhelpful* assessment of the situation can work powerfully in everyone's favour.

The Situation

You've had a bad day and feel fed up, so you go out shopping. As you walk down the road, someone you know walks by and, apparently, ignores you.

This triggers a cascade of responses, listed below – unhelpful on the left and reframed as helpful on the right. It's the same scenario but the information has been reinterpreted/reframed in a more positive light:

	Unhelpful	Helpful
Thoughts:	He/she ignored me – they don't like me	He/she looks a bit wrapped up in themselves – I wonder if there's something wrong?
Feelings:	Low, sad and rejected	Concerned for the other person, positive
Physical:	Stomach cramps, low energy, nausea	None – feel comfortable
Action:	Go home and avoid them	Get in touch to make sure they're OK

The same situation has led to two very different results, depending on how the situation is framed. But in both, how you think affects how you feel and what you do. In the example in the left-hand column, you've jumped to a conclusion without very much evidence for it and this matters, because it has led to you:

⊙ having a number of uncomfortable feelings
⊙ behaving in a way that makes you feel worse

If you go home feeling depressed, you'll probably brood on what has happened and feel worse. If you get in touch with the other person, there's a good chance you'll feel better about yourself. (It's also worth noting the element of reciprocity bias – discussed in Foundation Stone Four – present throughout the 'Helpful' column, which demonstrates that the more you exhibit concern for the other person's well-being, the better you feel about yourself.)

In their book *Winning Decisions: Getting It Right the First Time*, management experts J. Edward Russo and Paul J. H. Schoemaker provide a useful story to illustrate the power of framing.[40] A Jesuit and a Franciscan were seeking permission from their spiritual superiors to be allowed to smoke while they prayed. The Jesuit asked first whether it was acceptable for him to smoke while he prayed. His request was denied. The Franciscan reframed the question: 'In moments of human weakness, when I smoke, may I also pray?' and was granted permission.

Jordan Scott, a Canadian poet who stutters, was researching his theory that stutterers are often thought to be lying by those in authority. Having studied transcripts from the US military prison at Guantanamo Bay, he noted that interrogators there interpreted stuttering as evidence of dishonesty. He applied to visit Guantanamo prison, hoping to interview interrogators and translators; knowing this would be difficult, he applied for the visit as a poet rather than explaining about his research, so as to appear less threatening. He was told that, as a condition of acceptance of his visit request, he would not be able to interview anyone or record unauthorised human voices. He asked if, instead, he could record ambient noises around him. This reframing of purpose worked a kind of magic. Scott explains: 'The whole place is totally restricted. You can't go anywhere without an escort. You can't take any pictures. You can't talk to people. But this is the thing: they did not know what to do with me. I would say, "Can I go to Camp Iguana?" and they would say, "Absolutely NOT," and I said, "Well, I just want to do ambient sound," and they were like, "Oh. OK. I don't see an issue with that."'[41]

Framing potential outcomes as gains or losses can influence people's choices. Psychologists Amos Tversky and Daniel Kahneman were the first to demonstrate how preferences can change depending on how the same problem is presented.[42] In one study, a sample of doctors was asked: 'Imagine that the US is preparing for the outbreak of an unusual Asian disease, which is expected to kill 600 people. Two alternative decision frames to combat the disease have been proposed.'

Frame 1

In a group of 600 people:
- If Programme A is adopted: 200 people will be saved.
- If Programme B is adopted: There is a one-third probability that 600 people will be saved, and a two-thirds probability that no one will be saved.

Seventy-two per cent of doctors preferred Programme A; the remaining 28 per cent opted for Programme B.

Frame 2

In a group of 600 people:
- If Programme C is adopted: 400 people will die.
- If Programme D is adopted: There is a one-third probability that nobody will die, and a two-thirds probability that 600 people will die.

In this decision frame, 78 per cent preferred Programme D,

with its one-third chance of no deaths, with the remaining 22 per cent opting for Programme C.

In each case, the framing caused doctors to seek out the option that offered the most optimistic-sounding outcome: they were choosing life – gain, if you like ('will be saved' and 'nobody will die'), over loss ('two-thirds probability that no one will be saved' and '400 people will die'). However, as you might have worked out, programmes A and C are identical, as are programmes B and D.

Here's another example of how changing the frame can alter things dramatically. In the 1980s, Coca-Cola was doing very well – in fact, it had a 45 per cent share of the total soft-drinks market and it was happy with that. But the then CEO of Coca-Cola, Roberto Goizueta, changed the frame and reference point for this 45 per cent share by asking three simple questions, of which question three was perhaps the most significant:

1. How much liquid does any individual consume in a day?
2. How many people are there in the world?
3. What percentage of the entire liquid market do we have?

When the answer to question three turned out to be a paltry 2 per cent, it wiped the satisfaction from marketers' faces. This was quickly followed by the realisation that this new data opened up an enormous opportunity for the company, leading to a period of extraordinary growth. Coca-Cola's market value grew from $4 billion in 1981 to roughly $150 billion in 1997.

Similar effects occur in less-controlled, everyday circumstances. For example, it sounds more positive to say that a new-product launch has a '1-in-10 chance of succeeding' than a '90 per cent chance of failing'. If people are rational, they should make the same choice in every situation in which the outcomes are identical. It shouldn't matter whether those outcomes are described as gains or losses, or successes or failures. But the choice establishes different frames, and decisions may differ as a result.

Here are two anecdotes to illustrate the benefits of framing in everyday life:

⊙ Each year, some friends ran a sweet stall at an annual charity fundraising sale at school, and one year, along with the mixed bags of penny chews, they sold Mars bars. The Mars bars cost 50p each. For some reason, they didn't sell as well as the mixed bags of sweets at £1. So, the sellers took to saying 'It's 50p for one Mars bar or £5 for ten' and sold out in minutes. They'd *reframed* the sale in the language of a deal, suggesting the 'gain' of ten Mars bars was an irresistible bargain – and none of the children (or their parents) bothered to do the maths.

⊙ A friend described how on a trip to Florence with a boyfriend years ago, she accidentally trod in dog poo on leaving a church where they had been looking at frescoes. As she bent to clean it off, a pigeon followed suit on her head. 'Wow, that's meant to be really lucky,' said the boyfriend admiringly. See? Reframing.

When you're presenting others with choices and you want them to pick a particular option, try to make sure you express your preferred option using gain-related vocabulary, while framing the other options as losses. For example, saying 'If we set off at 5 a.m. we'll miss the traffic and have the whole day at the beach' is a gain frame, while 'We can leave as late as you like, but we'll end up sitting in traffic' is a loss frame. Here's another example of gain-framing language: 'Eat healthy foods and lower your cholesterol', where loss-framing might express the same principle as 'Eat healthy foods or lose years off your life.' Scientific studies have shown that gain-framing can be more effective in prompting preventative measures, as in 'Wear sunscreen and lower your risk of skin cancer', while loss-framing seems to work better for detection behaviour:

'If you fail to get regular mammograms, you are less likely to detect breast cancer when it may still be curable.'[43]

In fact, you will find it deeply rewarding to spend a little more time thinking about how you code or decode a piece of information. In our house, the line 'Let's reframe that' has become a staple nudge. Here are some examples of how you might reframe a negative situation to bring out a shiny silver lining:

- ⊙ You planned to meet a friend after work, but they cancel at the last minute – reframe disappointment as the gift of a night to yourself.
- ⊙ It rains every day on your long-planned weekend break – reframe miserable wet walks as bracing and invigorating.
- ⊙ Your flight is cancelled due to bad weather and you're forced to go home – reframe a lost trip and think about all the money you've saved.
- ⊙ You're sure you had £30 in your pocket when you left the house, but when you reach for it to pay for groceries, there's just an empty pocket – reframe the self-flagellation by imagining the delight of the person who picked it up.
- ⊙ You get the picture.

Chunking OR *How to eat an elephant? One bite at a time*

Complex tasks can seem daunting, even unachievable. In fact, if we look at a task in its entirety, we'll sometimes take the easy way out – 'It will take too long,' we tell ourselves; 'It will be impossible to do in the time I have'; 'I just can't face it right now'; and the procrastinating 'I'll do it tomorrow' as we hand on the responsibility to our future selves.

However, many studies have shown that if we break up big tasks into smaller pieces it can completely change our attitude to their achievability. 'Chunking' is a very simple method of playing to how our brains are wired and it's another element of reframing. Think back to your childhood and your parents' encouragement to eat just ten more peas (and then another ten) or just two more pieces of carrot (and then another two), and see how long chunking has been a part of your life! If you take a moment to break down a task into one step at a time, it's usually achievable. And taking things stage by stage also allows you to take a breath and tackle the task more calmly.

You already do loads of chunking; think about phone numbers and how you break them down into groups of numbers – you almost certainly give people your mobile number first as a group of four or five numbers, then the next three and then the last three or four. It's incredibly unlikely that you reel it off as a list of eleven numbers, and if anyone reads it back to you in any format other than 5-3-3 or 4-3-4, you probably won't recognise it! And, of course, you are being 'chunked' by others in different ways, too: when you phone a large retailer or utility provider and the automated response apologises for the wait, they're exceptionally busy etc., but you are sixth in the queue; and then, a little later, how sorry they are, they are exceptionally busy etc., but you are *fifth* in the queue. They are chunking you along your journey, showing you the little bits of progress you are making towards your goal, one step at a time.

47

Foundation Stone Four: Behavioural biases

Behavioural biases are predispositions that are wired into our brains and encourage us to behave in one way vs another – often irrationally.

There are probably more biases being identified all the time, but Wikipedia lists 185 and that's a lot to be going on with. We've identified the influences of the behavioural science Foundation Stones One to Three (Running on autopilot, System One and Two thinking and Context is ALL, respectively), and now Foundation Stone four kicks in, with a complex array of behavioural biases. We'll look at some of the most impactful ones here.

Reciprocity Bias: Do as you would be done by, with icing on the top

Our brains are wired with the tendency to behave towards other people in the same way that they behave to us. This is known as the reciprocity bias.

We think of it as a purely positive impulse, but it can work both ways. It is equally powerful used negatively. Ernst Fehr, Professor of Microeconomics and Experimental Economic Research at Zurich University, and Simon Gächter, Professor of the Psychology of Economic Decision Making at the University of Nottingham, put it like this in their article 'Fairness and Retaliation':

> A longstanding tradition in economics views human beings as exclusively self-interested … However, many people deviate from purely self-interested behaviour in a reciprocal manner. Reciprocity means that in response to friendly actions, people are frequently much nicer and much more cooperative than predicted by the self-interest model; conversely, in response to hostile actions they are frequently much more nasty and even brutal.[44]

Every positive or negative act you carry out, however small, can generate a ripple effect as you stimulate the reciprocity bias in the next person. The added advantage of any positive actions we initiate is that we get a kick out of them too – it's indisputable fact: our brain activity shows this to be so. Jamil Zaki, professor of psychology at Stanford University, suggests kindness could be a kind of 'psychological chocolate' in that 'people might actually enjoy doing kind things for others, and that might be an emotional engine for driving pro-social behaviours'.[45] His research tested the theory that reciprocity is intrinsically rewarding. Participants made decisions about giving away amounts of money to strangers while their brain activity was observed in an fMRI scanner. Results showed that when participants acted 'equitably' and gave money away without hope of recompense or thanks, the area of the brain associated with 'assessing rewards' was activated.[46]

Think about what happens when you let a driver out of a side road or greet a colleague with a big smile and a cheery 'Good morning!' and compare that with the day that begins with you slamming the front door in a huff and deliberately driving your car to block traffic at a junction.

In a poem called 'Small Kindnesses', Californian poet Danusha Laméris expresses perfectly the human need to exist in a kind world, suggesting that the impact of small gestures of generosity towards others might hold more power than we imagine:

> What if they are the true dwelling of the holy, these fleeting temples we make together when we say, 'Here, have my seat,' 'Go ahead – you first,' 'I like your hat.'[47]

Consider how your actions can ripple on through each encounter you have, good or bad. Consider how many positive, pleasure-giving vibes your actions could trigger as the day goes on, or how many unsettling negative vibes have prompted discord somewhere down the line.

Status Quo Bias: Why the 'same old, same old' keeps getting our vote

Our brains are wired with an in-built bias to avoid change; it means we prefer to stick with things the way they are and it can be responsible for encouraging us to avoid change wherever possible.

So, most of us like things to stay the same – we know where we are then, and there are no nasty surprises. It's standard practice to have both a conscious and subconscious tendency to try to keep things just the way they are. This brain wiring is aptly called the status quo bias. What this means is that we are subconsciously primed or predisposed to reject ideas, actions and suggestions (however big or small) that would disrupt the familiar in our daily lives and status. In fact, it's a bias that impacts on all aspects of our lives, from how we live day-to-day to how we spend our money or express our political affiliations.

Sit down, you're rocking the boat

We are wired with a tendency to stick to the status quo, to avoid rocking the boat and to put off doing things that are just not part of our everyday plan. If you have ever been a member of a close-knit group of friends or colleagues (or a book group, for example), and one of the other members suggests inviting someone else to be a part of that group, it's more than likely your instinctive response will be resistant. That's the status quo bias at work and it's likely to be based on the fear that a newcomer (a boat-rocker) might destabilise the group. After all, why take the risk if the group works well as it is? Of course, the flip side is you don't know what you're missing out on if you don't (at least occasionally) open up to a little new energy and excitement.

Sticking with our own default factory settings

Our attitudes predispose us to go with the status quo, and this affects our response to how 'choices' are presented to us. If a choice has already been made for us – the default option – we'll probably stick with it. This default setting is another aspect of the status quo bias. We have our own innate default settings that we have fixed fast to our sense of self – our tendency to say an immediate 'Yes' or 'No' to invitations, for instance; our stated dislike of big concerts or small ones, pop music or classical music; our refusal to ever contemplate eating broccoli or Brussels sprouts; our habit of always taking the same seat in class or on the bus – it can be as simple as that. It's all part of our default scaffolding. But it doesn't always have to be that way. We can change our defaults or simply reset them for a time – and see what happens.

If we leave the defaults in place, we might also be playing with fire. Think about things like phones and computers, which also have default settings that most of us do nothing to change. And think about how the UK phone-hacking scandal in 2011 was possible because journalists and private investigators used the default-set PINs offered by mobile phone companies (usually 1234, 0000 or 3333) to access the messages on their targets' phones. Where targets had changed the default PIN to something more individual, hacking was not possible.

We all know the tick-boxes that show up on online checkout pages and that will condemn us to communications from an indefinite number of third-party companies if we fail to change the default to *un*ticked – or, indeed, ticked: sometimes they play around with that, so you need to keep on your guard.

There is plenty of good, solid evidence to prove that when it comes to changing our utility suppliers, our jobs or our attitude to investment, and so on, we have a marked tendency to hold fast to the incumbent; to stick with the status quo. The moral of this bias is to question who the status quo/default setting serves best. If it isn't you, then rock your boat or change the default.

Using the default setting for manifesting good

An understanding of our predisposition for the default can be used to nudge or steer us into better behaviours, such as this one initiated by Starbucks: in a rocking-the-boat manoeuvre, the company effectively 'punished' its customers for sticking with the default in an attempt to encourage them to change their habits. In February 2018 Starbucks initiated a trial levy on disposable cups that saw 5p added to the cost of each coffee in one of their takeaway cups. Their aim was to nudge customers to make a more environmentally sound choice. The 5p levy aimed to shift customers from their default setting (that of taking a coffee in a disposable cup), to encourage new behaviour (bringing a reusable cup), reminiscent of the supermarket plastic-bag toll. This is from the Starbucks website at the time:

> The trial, which will see 5p automatically added on to any paper cup purchased, will last three months and net proceeds from the cup charge will be donated to environmental charity and behaviour change experts, Hubbub. Together, we'll evaluate the results of the trial and the impact on customer behaviour, helping to reduce waste and encourage the use of reusable cups.

(Note the presence of behaviour-change experts Hubbub here.) Starbucks once also offered a 25p discount if you brought your own reusable cup – channelling reciprocity bias and change-the-default running in parallel. That's two experiments in behavioural change conducted right before your eyes.

In contrast, the status quo or defaults around a particular subject, such as body shape, need to be challenged more directly.

Time to change the default on body image?

Default settings can be imposed by societal attitudinal norms. We find ourselves thinking something is 'right' because it's what society and the media present as the ideal. One way of exploring this is looking at 'ideal' body shapes through even relatively recent history. Victorian women in Britain aspired to a plump, ample-bosomed figure, albeit with a teeny waist (pulled in tight by corsetry rather than abs). Then, in the 1920s, a boyish physique was fashionable and chests were flat. Bodies got curvy again in the 1930s through to the 1950s before slimming right back down in the 1960s and heading, via the supermodels of the 1980s (curvy again, with very long legs), to what became known as 'heroin chic' in the 1990s (a style whose identifying features – dark-ringed eyes, waif-like thinness and translucently pale skin – were associated with excessive drug-taking). The journey is something of a roller-coaster ride – and it is certain to keep on rolling.

For now, it seems the tide has turned against using painfully thin models on the catwalk, and health and strength are gaining momentum. But most women would probably still admit to considering a thin body as a better body and probably the body they aspire to. This unappealing twenty-first-century ideology has become deep-rooted in most women's brains as the default setting. When will we reach a tipping point of 'Are you beach-ready?' articles, shake off the belief that 'Nothing tastes as good as skinny feels'?[48] and move into a new era? To reset the default we need more focus on the benefits of health and strength.

Normalising Bias: Why we're attracted to shiny new stuff

As if our propensity for automatic actions wasn't enough already, our brains are wired to 'normalise' things – a process

that's part of adaptation and learning. (This bias is also known as 'hedonic adaptation'.)

Normalising bias is when what was once fresh, new and exciting becomes the norm. New skills, new possessions, new friends, new experiences, even new relationships – all can fall victim to this bias. Once we have adapted to something, we no longer need to apply the level of conscious attention that a new thing demands because, from our brain's perspective, we don't want to waste unnecessary time '*re*learning' something we already know. So, we become accustomed to possessions, people and experiences, gradually our sharp focus on them diminishes. They slip from front-of-mind newness and the associated high-emotional energy that was initially attached to them and become part of our everyday, part of our 'normal', operating more and more in our subconscious and less and less in our active consciousness. Normalising bias is also hardwired into us via social and peer-group norms as we quest for the next exciting experience, the next new thing.

As the US behavioural economist Tibor Scitovsky put it in his 1976 book *The Joyless Economy: The Psychology of Human Satisfaction*, 'We consume and experience pleasure, as long as what we consume is novel.' What is new and thrillingly pleasurable one day becomes next week's comfy slippers. Comfort is nice but it's not exciting or memorable; it makes the world a flatter, softer place and it doesn't move us forward or encourage us to develop.

If we don't want everything to become 'same old, same old', we need actively to fight the normalising, or hedonic adaptation, process of the brain. If we work on thinking about this bias in action in our lives, we can occasionally increase levels of intensity in our response to the world around us.

Social norms: We all like to follow the herd and we'll jump on the bandwagon given half the chance

There are two main types of social norms: descriptive and

injunctive. Let's look at each one separately and learn how you can recognise them when they're at work.

Descriptive social norms

This is where we have a common tendency to adopt the opinions and follow the behaviours of the majority; to do what other people around us are doing. (It's also called conformity bias, which spells it out perfectly.)

Descriptive social norms encourage us not to stand out, not to appear different, and to trust to the whim of the majority. They are particularly in evidence in the way we choose to present ourselves to the world – our choice of clothes (we like to be fashionable), the films and TV programmes we watch (other people have praised them), the technology we own (we like to keep up to date). It's a tendency that has been wired into us for millennia and is probably a survival instinct from a time when standing out, or not doing what others did, was likely to be a quick recipe for early extinction. We all know from nature documentaries what happens when predators manage to cut one animal off from the safety of the herd.

Here's Loretta Graziano Breuning, Professor Emerita of Management at California State University, on how this works on a purely biological and instinctive basis:

> Your mammalian brain evolved to seek the safety of the herd. Mammals release large quantities of oxytocin at birth. That bonds an infant to its mother, which protects it from wandering off and getting eaten. Gradually, the mammal brain wires itself to feel attachment to a herd or pack or troop rather than just its mother. A herd animal releases cortisol when it can't see at least one other herd-mate. But some herd-mates are more trustworthy than others. Stronger gazelles push their way to the center of the herd where it's safer from predators, leaving the weak ones exposed to predators at the fringes.

But herd animals stick with their herd regardless, because they get eaten by a predator if they leave. Herd animals don't rehash this decision all the time. They just respond to the neurochemistry that makes familiars feel good and isolation feel bad.[49]

Our contemporary sense (evolved from the playground, perhaps) that all is not well when we look or behave too markedly differently from other people has this same hint of evolutionary psychology about it. It is an instinct that has developed to compel us to respond favourably to impulses like 'All for one and one for all' and 'I'll have what she's having.' And it's an instinct that serves us well. It usually makes sense to emulate what other people are doing (especially if they are all running in one direction and screaming in terror). Scientific studies have shown that it is in fact our brains that impel us to copy others. The bottom line is, we feel slightly uncomfortable if we don't.

Robert Cialdini, Emeritus Professor of Psychology and Marketing at Arizona State University, commented on a Chinese study of social norms in which a restaurant owner was reported to have put 'These are our most popular items' against particular dishes, which immediately became 17–20 per cent more popular![50] Other people's recommendations offer us a shortcut to decision-making. They save us the effort of reasoning our way through a decision on our own.

A friend of the authors had to have an endoscopy, a procedure in which an endoscope – a long, thin, flexible tube with a light and camera on the end – was introduced through her mouth to her oesophagus and then her stomach. The medical staff offered her sedation before the procedure began, to make it more comfortable. She asked whether other people having the procedure typically required sedation and was told that '90 per cent of people have the procedure without sedation and are fine'. So, she went without sedation too – but she wished she hadn't. Sometimes we need to fly in the face of social norms.

In the early 1950s, many psychologists were prompted to try to understand why so many Germans accepted Hitler's racist and anti-Semitic vision without question. One of the most famous experiments was conducted by a Gestalt psychologist called Solomon Asch.[51] He recruited volunteers to participate in what he called a vision test. Seated in a classroom with other participants (who, unknown to the volunteers, were all active players or confederates in the experiment who had been given a script), they were presented with two cards like this:

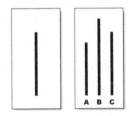

They were then asked to identify which line on the second card was the same length as the single line on the first card. When left to themselves, volunteers picked line C. However, if the confederates answered first and insisted that B was the correct answer, 33 per cent of the volunteers overrode their instincts and agreed.

More recently, fMRI technology – where scanners can detect which brain regions are active, with areas of the scan lighting up when people carry out different mental tasks – has allowed the reason for this malleability to become clear. In 2005 a US research team led by neuroeconomics and psychiatry professor Gregory Berns conducted fMRI scans on participants as they performed 3D-image-rotation tasks in a similar setting to Asch's study, so they could analyse what was occurring in participants' brains when they conformed or went against the group.[52] This is what they found:

⊙ When participants conformed to the group and gave the incorrect answer, brain images showed that their entire perception changed: they saw what others saw.

⊙ When participants went against the group, scans showed that the emotional part of the brain was activated, caused by the pain of going against the group.

As in Asch's study, participants conformed to the confederate majority view 41 per cent of the time, on average, despite having made a different choice when working independently.

From this brain-imaging work, Berns found that when people conform in experiments like Solomon Asch's they actually see the situation as everyone else appears to. If the group insists a circle is a square, for example, the conforming participants will literally *see* a square; they aren't lying or trying to please – they are conforming. Berns found that the conformity was associated with functional changes in the brain: participants' visual perceptions were shown to have altered. They also seemed to think less – there was less activity in parts of the brain we associate with System Two thinking. When they went against the views of the group, they experienced emotional anxiety from not agreeing and having independent views. Berns' research found that brain mechanisms associated with fear and anxiety play a part in situations where a person feels his or her opinion goes against the grain. Non-conformity, or independence from the group, was associated with increased activity in the brain's amygdala and caudate nucleus, indicating emotional load. What the experiment didn't identify was how the non-conformists managed to resist the impulse to conform. How are some people content to stand out from the group, given the pain of doing so?

In a later study conducted by the Rotterdam School of Management at Erasmus University, findings showed that when people hold an opinion differing from others' in a group, their brains produce an 'error' signal.[53] When this error signal is strong the reward signal slows down, and this imbalance indicates what people probably perceive as the most fundamental social mistake – that of being too different from other people. And if you really want it spelled out, how about this forbidding-sounding statement from the Erasmus University team: 'Deviation from the group opinion is regarded by the brain as a punishment.'

In this study, the participants (who were all female) had to rate 222 faces on their physical beauty and give them marks on a scale from 1 to 8. After having rated a face, each subject was shown a higher, lower or equal mark and was told that this was the average score awarded for that face by others taking part in the experiment. On completion of the task, participants and researchers chatted over refreshments. Unexpectedly for the study participants, the study leader then asked them to evaluate the physical beauty of the same faces once again.

The outcome of this second face-assessment was that most of the subjects shifted their opinions towards the average opinion of others that they had been told about earlier. Although the degree to which the participants changed their judgement varied considerably, no one proved to be totally immune to the group pressure. Everyone made at the very least minor changes in the direction of the group's opinion. The researchers used MRI scanning technology to show the brain activity triggered by this study and concluded that when we find that what we think is out of line with what others around us think, our brains respond by suggesting we have made a mistake and attempt to readjust our thinking so that we might well find ourselves 'changing our minds' to agree with everyone else.

Other studies have shown that we will do what others are doing even when there is no rhyme or reason for their actions. There are hilarious videos of people sitting in crowded waiting rooms where, when a bell rings, the other people waiting all stand up.[54] After a couple of repetitions of this scenario, the unwary newcomer stands too, and continues to stand each time the bell rings. In another study people entering a lift will eventually copy the other people already in it whenever they (incomprehensibly) turn to face the back.[55] It's funny to watch, and we all assume we ourselves would rise above it, be smarter than the rest. But think about the last time you saw a queue of people in the street and how you immediately wanted to know why they were queuing and were somehow tempted to join the back of the line; and about how, if you were to see a crowd of people looking up towards the sky, it would take considerable effort not to follow suit.

In simple terms, our brains want us to do what others are doing, because if a lot of other people are doing something, it might make sense for us to do it too.

Amazon plays to this descriptive social norm extremely well, using prompts and nudges like 'People who bought this also bought this' to encourage us to join the herd. The success of *Fifty Shades of Grey*, the soft-porn literary phenomenon of 2011, is a fine example of the influence of the herd instinct in action: sales rocketed when the press flagged the success of the book and delivered a big 'Everyone else is reading it so it's OK for you to read this book too' permission-to-purchase nudge to other potential readers.

Waitrose gives shoppers a green plastic token at the checkout and encourages them to select one of three local charities to be supported by dropping the token into a clear plastic case labelled with that charity's name. The different levels of selection are very obvious – you can see who has more or less – and a clear 'winner' usually emerges – a 'winner' whose success is compounded because the more tokens it has, the more it will continue to attract, on the basis that shoppers are reassured by herd instinct that they are selecting the 'right' one. As one shopper was heard to say as she dropped her token into case for the most popular cause: 'I think this one must be the best, because it's the one most people have chosen!'

In their paper 'The Unbearable Automaticity of Being', psychologists John Bargh and Tanya Chartrand write that when we do what others are doing, it does more than simply make us feel comfortably part of the group, it also 'increases liking and creates a sense of smooth interactions':

> Within a social group setting, one is more likely to get along harmoniously with others in the group if one is behaving similarly to them, compared with being 'out of sync' and behaving differently. Thus it makes sense for the default behavioural tendency in an interaction to be based on one's perception of what the other person is doing.[56]

Injunctive social norms

Injunctive social norms are what we *perceive* as being approved of by others; as conveying societal approval and acceptance.

We often do things because we think it's what society expects of us. This works well for actions like giving to charity and putting our litter in the bin, but sometimes we can over- or underestimate society's views.

Writing in the *New York Times*, columnist David Leonhardt focuses on two dangerous misperceptions in American life. Each is a great example of an injunctive social norm that negatively impacts people's lives, one with potentially fatal consequences.[57] The first is the concept that the flu vaccine is relevant only for the elderly or young children and that other people don't need one. 'But none of that is correct,' he says. 'You should get a flu vaccine every year – for your own sake and for the sake of public health.' The second is the perception among those on low and middle incomes that college costs are sky high and completely unaffordable, at tens of thousands of dollars. However, Leonhardt points out that 'average in-state tuition at public colleges will be just $4,140 this year'. Correcting injunctive social norms like these can change lives.

Authority Bias: The power of status, white coats and uniforms

This relates to the tendency to alter our opinions or behaviours to those of someone we consider to be an authority on the subject. We use the authority view as a mental shortcut.

We haven't got much of a chance against this one as we move from one authority figure to another throughout our lives. We start off as children, with our all-seeing, all-knowing parents, grandparents and other adult authority figures. (And who doesn't remember a teacher whose voice could instil fear, even now?) Then we move on through life with employers, police officers, doctors, traffic wardens – a plethora of people, often in uniform, telling us what to do. Uniforms are authority-signifiers, and people in uniform offer us an instant visual anchor that can nudge our behaviour.

Uniform is a shortcut that communicates authority, trust, official status, permission, warning, etc. In hospitals, white coats* and scrubs routinely engender trust, reliance, dependability and hope among patients and visitors alike. Police officers, firefighters, doctors and ticket inspectors are all people whose uniforms instantly signal their authority over us, and, to some degree, their trustworthiness too. Uniforms in the right context tell us that the people wearing the uniform are in charge and that they know what they are doing. But authority can trick us, too. We've all seen the movies where the bank robbers don the uniform of security guards and make off with the loot.

---

* Not to be confused with 'white coat syndrome', which causes some people to exhibit symptoms of raised blood pressure in a clinical setting – though not specifically when confronted by a doctor in a white coat!

In 2009 a team of psychologists from the US and the Netherlands devised a study using fMRI scans to show how the perceived expertise of others can cause the decision-making areas of our brains to shut down as we submit to their authority.[58] Gregory Berns led the study in which participants were asked whether they would swap a guaranteed payment in favour of a lottery that would deliver a higher payout if they were to win. The control group made the decision for themselves while the test group received advice from a financial expert. Those in the control group showed activity in brain regions associated with making decisions and calculating probabilities. In contrast, brain activity in those regions flatlined in test-group participants who received advice from an expert. We effectively relax our conscious effort when in the presence of those we believe to be better qualified than ourselves.

As Berns pointed out, 'This study indicates that the brain relinquishes responsibility when a trusted authority provides expertise. The problem with this tendency is that it can work to a person's detriment if the trusted source turns out to be incompetent or corrupt.'

Have you ever seen the 1940s advertising slogan 'More Doctors smoke Camels than any other cigarette' – a cunning use of authority bias complete with a picture of a trustworthy-looking chap sporting the regulation white coat, cigarette in hand? Rothmans did the same thing, with ads showing a male hand variously on the wheel, gearstick or door handle of an expensive-looking car, wearing an expensive-looking watch and holding a cigarette. The hint of gold braid on the man's cuff implied that he was a pilot or a naval officer. The ad's strapline was '*...when you know what you're doing*'. Authority bias loud and clear. It's perhaps interesting that e-cigarettes might be said to be receiving twenty-first-century medical endorsement, with Public Health England suggesting they could be prescribed to patients wishing to quit smoking.[59]

Advertisers have been using the authority bias on us for years – '*Mother knows best*' (Mother's Pride bread, c. 1970), '*Eight out of ten cats prefer...*' (Whiskas, c. 1975); they associate supermodels with fashion brands, cosmetics and perfumes to exploit the 'authority' of their beauty/physical perfection. I remember a law firm that used as their TV talking head an actor who had played a hard-bitten detective in a TV cop show called *The Bill*. This was a prime example of authority creep – they exploited the fictional authority of his character to boost their credentials.

Hormone Replacement Therapy – the pros and cons – is a newsworthy subject. Views differ. Do the risks outweigh the benefits? Do the benefits outweigh the risks? And those in 'authority' don't always agree. An article in *The Lancet*,[60] reported in *The Times*,[61] suggested that HRT was linked to more breast cancers than had been thought. Cause for immediate concern? Read on. The *Times* article ends with information from the vice president of the Royal College of Obstetricians and Gynaecologists that 'Most female gynaecologists take HRT, most of the wives of male gynaecologists take it.' What on earth do you do when there's a *conflict* of authority bias?

Of course, where genuine authority and expertise exist, there is no doubt that they exert a powerful force. Sometimes, of course, we just enjoy the belief in the magic the authority figures promise.

Egocentric Bias: Let's talk about me, me, me!

We are, in essence, each wired to make the world revolve around us, and have a tendency to view ourselves and our actions through rose-tinted glasses.

This means we view our behaviour, our role in a discussion or our contribution to something – anything – more positively than it might deserve. And it is the egocentric bias that causes us to regard ourselves as being 'right' in pretty much every circumstance.

We also have a tendency, when remembering past events, to exaggerate our roles in them – the fish we tell you we caught was way bigger than the fish we *actually* caught (yes, that old chestnut is caused by the way our brains are wired), we ran fastest, did most of the heavy lifting taking that wardrobe upstairs, won the argument, were the pivotal member of the team. If, after a meeting, you were to ask participants to rate their input in percentage terms, there's no doubt the total would add up to more than 100 per cent! How much housework do you think you do, compared with your partner? That's a classic example of how egocentric bias operates. Comedian Robert Webb expresses it perfectly:

> Compared to my wife, I think I do most of the shopping and more cooking of the kids' teas. Loading and unloading the dishwasher is pretty even – by which I mean I think I do more. I do slightly less of the laundry, which probably means I do almost none of it. You see how this works?[62]

Closely associated with the egocentric bias is the self-serving bias – the tendency to claim more responsibility for our successes than for our failures. It follows on adroitly from the egocentric bias because, once again, it's all about *us*. It also describes our tendency to evaluate ambiguous information in a way that's beneficial to our interests – that bears out what we want to believe.

In a 2005 cartoon in her 'Tottering-By-Gently' series, Annie Tempest depicts an elderly couple relaxing on the sofa: 'When one of us dies, I'm going to go and live in the South of France,' muses the husband.

How's this for self-serving? Experiments have shown that just about all of us believe ourselves to be:
⊙ more competent than our co-workers
⊙ more ethical than our friends
⊙ friendlier than the general public

⊙ more intelligent than our peers

⊙ more attractive than the average person

⊙ less prejudiced than other people in our local area

⊙ younger-looking than people the same age

⊙ better drivers than most people we know*

⊙ better behaved/more successful than our siblings

⊙ likely to live longer than the average lifespan?[63]

How would you judge yourself on these dimensions? A good demonstration of egocentric bias, and its ability to allow us always to find space for ourselves on life's winners' podium, is a study in which a class of computer science students answered a survey after getting the results of their first test. Those who had passed the test said they thought being computer literate was an important skill in today's changing world. Those who had failed the test said computers were for geeks.[64]

Dunning-Kruger bias OR Illusory superiority
We can overestimate how good we are at things we don't really know how to do.

We are all prey to the influence of the superiority bias, also known as the Dunning-Kruger bias. It makes us overestimate our skills and abilities while at the same time endowing us with a total lack of self-awareness that we're doing it. Two psychologists from Cornell University, Justin Kruger and David Dunning, identified the bias when in 1999 they ran an experiment to see if people who lack the skills for something also lack the awareness of their lack of ability.[65] Their results revealed that it's more than likely that if you can't do something, you also

---

*   In studies carried out for Tali Sharot's book *The Optimism Bias: A Tour of the Irrationally Positive Brain* (Pantheon, 2011), 93 per cent of those surveyed believed themselves to be better drivers than over half of all other drivers on the road.

lack the skill it takes to do it, and yet you still think you might be pretty good at it. Here's a real-life example. In an art gallery, a couple are watching a film in one of those dark, curtained rooms off the main display area. The film shows a naked man dancing and twirling his genitalia, rather joyfully, round and round. The watching man turns and scoffs to his wife: 'I could do that…' She stops him dead: 'But you didn't.'

Is it a surprise to learn that there's a gender bias at play here? Studies show that men are more likely to overestimate their intelligence while women underestimate theirs. Men's confidence in their ability works for them too; it helps others to believe in them. In 2003, David Dunning and the Washington State University psychologist Joyce Ehrlinger examined the relationship between confidence and competence.[66] They gave male and female college students a quiz on scientific reasoning. Before taking the quiz, the students rated their own scientific skills. The psychologists wanted to cross-reference students' perceptions of how scientifically skilled they were with their confidence about the correctness of their answers to the quiz – two very different things.

The women rated themselves more negatively than the men did on scientific ability, giving themselves an average of 6.5 on a scale of 1 to 10, while the men gave themselves 7.6. When it came to assessing how well they had answered the questions, the women thought they had got 5.8 out of 10 questions right; the men a more confident 7.1. Ehrlinger and Dunning followed up the test by inviting all the students to attend a science competition (with prizes). Though none of the students knew how they had performed in the test, only 49 per cent of the women signed up, compared with 71 per cent of the men. The implications of the findings are related to women's career progression, since women are less confident in their abilities, they don't push for more in the workplace. In their concluding remarks, Ehrlinger and Dunning suggest that when you need an accurate understanding of how well you are doing, ask an independent witness rather than relying on your own assessment. You'll be most likely to judge yourself as either better or worse than you actually are.

A 2018 BBC investigation into the salaries of top NHS consultants revealed a number of disparities that could well be linked to the Dunning-Kruger bias.[67] Of the hundred highest-paid doctors in the NHS, only five were women, and the highest-paid doctor, a man, earned nearly two-and-a-half times more than the highest-paid female doctor; on average, female doctors earned £14,000 less than their male counterparts. Higher levels of overtime accounted in part for the differences, but so did the greater likelihood that the men had applied for, and won, clinical excellence awards worth £77,000 a year. Former Co-chair of the NHS Consultants' Association Dr Jacky Davis says, 'In my experience, men are better at pushing for more money, putting the case for awards and they get them.'

Hillary Rodham Clinton wrote about this in her post-2016 election-analysis book *What Happened*:

> Over the years, I've hired and promoted a lot of young women and young men. Much of the time this is how it went:
>
> ME: I'd like you to take on a bigger role.
> YOUNG MAN: I'm thrilled. I'll do a great job. I won't let you down.
> YOUNG WOMAN: Are you sure I'm ready? I'm not sure. Maybe in a year?[68]

A female friend described a day-long training seminar for magistrates. Two thirds of the delegates were women. At one point during the session they were split into groups to tackle a problem case, after which one person from each group presented to the rest. In spite of being in the minority, in almost every case it was the men who volunteered to present. So, is there an associated confidence issue here?

In their book *The Confidence Code: The Science and Art of Self-Assurance – What Women Should Know*, Katty Kay and Claire Shipman discuss what lies behind the confidence gap, noting that women are caught in a catch-22 situation when confidence (read 'overconfidence') can get a woman labelled as aggressive or 'a bitch'.[69] Where women let themselves down, they conclude, it is in the choice they make 'not to try', to give up in anticipation of poor performance or failure. Try, fail, try again, fail better and above all keep at it and beware the Dunning-Kruger bias would seem to be the go-to advice here.

A recent thread on Twitter demonstrated this very issue.[70] Former BBC Radio 4 *Woman's Hour* presenter Jane Garvey asked the question: 'I wonder if anyone can think of a topical example of that well-worn cliché that a man will have a punt at a job for which he's woefully under-qualified, but a woman might be more inclined to hold back?' Of the many, many replies that flooded back, the following were particularly illuminating.

From Jess Phillips, Labour MP for Birmingham Yardley:

> In every advert for a job at Women's Aid we said explicitly (citing the relevant sections of the equality act) that you had to be a woman to apply. Every single time without fail, at least 2 men applied. Every single time.

And this response from an employer:

> I'm hiring for a software developer role at the moment and I've seen this over and over again. Men exaggerate their skills on their CV, women underplay it. Men talk well but their practical tests are poor. Men generally ask for 10–20% more money too.

Optimism Bias: Viewing the world through rose-tinted spectacles

Optimism bias, or overconfidence, is the tendency to be over-optimistic about the outcome of planned actions and our likelihood of experiencing good events, while underestimating the likelihood of suffering from negative ones.

In her book *The Optimism Bias: A Tour of the Irrationally Positive Brain*, Tali Sharot, a research fellow at University College London's Wellcome Centre for Human Neuroimaging, suggests the human brain is 'hardwired for hope'. She says our brain activity makes humans predisposed to be hopeful, even when this flies in the face of reason, circumstances or experience. And although it can sometimes mean we make miscalculations – when we don't save for the future or increase our pension contributions, when we assume we won't get sunburned and forego the sunscreen, when we fail to get regular health checks because 'we're fine' – according to Sharot, 'the bias also protects and inspires us: it keeps us moving forward rather than to the nearest high-rise ledge. Without optimism, our ancestors might never have ventured far from their tribes and we might all be cave-dwellers, still huddled together and dreaming of light and heat.'[71]

Even the most pessimistic people hope for the best while expecting the worst. We don't get married expecting to get divorced; we don't buy lottery tickets convinced that we won't win; we don't book holidays with the assumption that we'll have a horrible time; we don't go to restaurants anticipating ordering food we won't like. Our day-to-day plans – our whole life plans, come to that – are mostly devised with the prospect of favourable outcomes. Studies mentioned in Sharot's book show that most respondents vastly overestimated how long they would live. (Almost everyone believes they will outlive the average person.) One out of every ten people said they would live to be one hundred. In reality, two out of 1,000 people will live long enough to open a birthday card from the King.

Not surprisingly, optimism bias and overconfidence can cause problems in large-scale building plans. The UK government has gone so far as to implement warnings and instructions to its various departments, specifically highlighting the optimism-bias effect. Here's an excerpt about optimism bias from HM Treasury's *The Green Book: Central Government Guidance on Appraisal and Evaluation (2020)*:

> There is a demonstrated, systematic, tendency for project appraisers to be overly optimistic. To redress this tendency appraisers should make explicit, empirically based adjustments to the estimates of a project's costs, benefits, and duration … It is recommended that these adjustments be based on data from past projects or similar projects elsewhere, and adjusted for the unique characteristics of the project in hand. In the absence of a more specific evidence base, departments are encouraged to collect data to inform future estimates of optimism, and in the meantime use the best available data.[72]

It would have been wise for those responsible for budgeting for the rebuild of the US Embassy in London to have paid attention to these bias warnings. The new embassy (which relocated from Mayfair to Nine Elms in Battersea) was originally budgeted at $275 million, in 2008, but even by the time construction work began, in 2013, that budget had more than doubled, to $620 million.[73] The final cost in January 2018 was more than $1 billion.

Another way of managing investment and time-intensive projects is to apply the practice of a 'premortem' at the outset. Premortem is a technique devised by psychologist Gary Klein. It can also be described as prospective hindsight and is simply where you imagine yourself in the future and the project you are currently planning is over, having failed spectacularly. In Klein's scenario, everything has gone as badly as you feared. Now ask yourself: why?[74] Once you have imagined the worst, you can plan for it, putting structures in place to prevent bad stuff happening, or making it easier to deal with if it does. You can apply it to anything you're planning – from high-stress events, like weddings, to a day at the beach.

Our optimism when it comes to future plans (big and small) – exam results, good weather for a picnic, finding a parking space – is founded on our tendency to regard similar past experiences in an over-positive light (retrospective impact bias). We remember the good outcomes and forget or shed the bad ones. If we were to consider our future plans using only external evidence to guide us (discounting the validity of our own introspective thoughts and feelings), we would be much more likely to arrive at an accurate assessment of outcomes. You will perhaps recognise optimism bias from tasks you've undertaken that you thought would take only an hour or a two but took so much longer (flat-pack furniture, anyone?).

There's a cautionary note relating to optimism and overconfidence. We like it in other people and are inclined to warm to confident and optimistic people. We probably think we should be more like them. Leaders are expected to display these traits. But therein lies the potential for a misstep. Leaders are expected to be decisive; quick decision-making is associated with leadership, and we like our leaders to be John Wayne figures. In an interview with the management consultancy McKinsey & Company reported in March 2010, psychologists Gary Klein and Daniel Kahneman discussed the problems of an overreliance on overconfidence and intuition in leaders:

KLEIN: Society's epitome of credibility is John Wayne, who sizes up a situation and says, 'Here's what I'm going to do'– and you follow him. We both worry about leaders in complex situations who don't have enough experience, who are just going with their intuition and not monitoring it, not thinking about it.

KAHNEMAN: That's one of the real dangers of leader selection in many organizations: leaders are selected for overconfidence. We associate leadership with decisiveness. That perception of leadership pushes people to make decisions fairly quickly, lest they be seen as dithering and indecisive ... There's a cost to not being John Wayne, since there really is a strong expectation that leaders will be decisive and act quickly. We deeply want to be led by people who know what they're doing and who don't have to think about it too much.[75]

Discounting the Future Bias OR *The power of now* OR *Why it is easier to choose cake today*

We have an in-built wiring to discount the future in favour of the present moment. Cake today and a healthy salad tomorrow?

We don't like waiting and we don't like losing, whether it's waiting for rewards or losing out on returns on an investment – discounting the future bias and time-inconsistent preferences are both long-winded ways of describing our tendency to prefer immediate rewards to long-term future gains. (It's also known as present bias.) All the things we know would be 'good for us' in the future but will cause us some inconvenience or pain today – things like not having that second piece of cake or glass of wine, giving up smoking, getting fit, saving money or making bigger, or any, pension contributions – these are the things that are affected by this bias. We are more oriented towards short-term than long-term satisfaction, and that's why we find it hard to do any of them. J. D. Trout, a professor of philosophy at Loyola University in Chicago, expresses it perfectly when he says we are addicted to 'present pleasures': 'We are all too aware of our present desires: the pleasing anticipation of the taste of cheesecake, the long draw on the cigarette, basking on the beach, or sex.'[76]

It is hard to resist the siren call of gratification right *now* and to choose instead to postpone it until a *future* time. However, we are perfectly happy to commit our future selves to hardships of one kind or another – it's why, although we personally can't start the diet today, or go for a run just now, or forego this cigarette we're about to light, we're confident that the future us will be able to do it – tomorrow, on Monday, next week, sometime, anytime; just not now. Our tendency to discount the future is also the reason we might overcommit ourselves in the future – agreeing to do things then that we would never commit to do today. We haven't been able to exercise much this week, but next week we're really going to go all out. If you ask me for a two-hour meeting tomorrow afternoon, I'll most likely turn you down, but I'll happily agree to multiple meetings in a month's time. When that future date arrives, I'll seriously regret my busy schedule. How could I have done this to myself?

Ask us what we will have for lunch each day next week and we're likely to list a varied and healthful menu. When 'next week' comes around, our immediate lunchtime choices are likely to be much less varied and probably less healthful too. And our exercise plans? They'll probably go the same way. I'll definitely have that doughnut today while Future Me eats salad.

J. D. Trout suggests that our apparent frailty in the face of this bias can be bad for us, especially in the context of future financial planning. Even though we can discover for ourselves that to live comfortably in retirement we would need an income of around 80 per cent of our current pay, we still fail to put enough money into our pension pots. But help can be at hand, and, in the US at least, it is behavioural science that has driven the development of a pension product that attempts to tackle the problem. The Save More Tomorrow plan allows employees to divert a portion of their *future* salary increases towards retirement savings. The plan's genius lies in the upfront commitment it demands: getting prospective plan-holders to sign up to starting the plan in three months' time (i.e. committing their future selves to do it) and, further, that the additional contributions to the plan come directly from salary *increases*, meaning that individuals will never see the impact of the increased contributions on their take-home salary and so they won't have to regard the pension contributions as money lost.[*]

The concept of Future You is even being harnessed by advertisers. In an email from fashion retailer Boden, the subject line reads: 'Your future self will thank you.'

It can help to remember your future self when you are about to commit to something in the distant future. Instead of instantly agreeing, be kinder and just say 'No'. Your future self will almost certainly thank you.

Commitment Bias: If you commit to it, you're more likely to see it through

Simple small devices to commit ourselves to an action can help to put paid to procrastination, inertia or impulsiveness.

This is a useful one, so pay attention. You probably use it already without knowing that it's part of the behavioural science stable of biases. If you have a gym or jogging partner – that's commitment bias; if you write notes to yourself and leave them on the fridge to try to stop yourself from snacking – that's commitment bias. It's about building a structure around a plan of action that will help you to keep to the plan. So, if you're planning on training to run a marathon, starting an exercise regime or giving the status quo a run for its money, this is the one for you.

The insight behind commitment bias is that making a definite

---

[*] This concept was developed by Richard H. Thaler and Shlomo Benartzi. Over a twenty-eight-month period, savings rates tripled.

commitment to someone or something means you are more likely to follow through with a plan. It's why writing things down, telling others about your aims, leaving yourself no alternative other than to do what you have said you will do – even betting money against yourself – are all forms of commitment. Sites such as stickK.com promote self-contracting, which taps into this bias by asking people to bet money or some item of value against their goals.

Here is a great and simple example of applying the commitment bias to students suffering from study procrastination. Dissertation Write-In is a study programme run each quarter by Graduate Student Affairs at the University of Chicago.[77] Twenty graduate students pay a $50 deposit at the start of the week to have desk space in the library for a set number of hours per week, and commit to showing up at the library every day at 8:30 a.m. At the end of the Write-In, if they've attended every session, they get their money back and a 'completion' T-shirt that reads: *'Scribo, ergo conficiam'* – or 'I write, therefore I finish.' The study programme includes free coffee, snacks and lunch, and has proved very popular. Graduates find the structure and neutral location to be extremely helpful commitment strategies. There's no free internet access; nothing to pull their attention away from their goals. And seeing other people also hard at work helps them to do likewise. Productivity improves and the group cohesion encourages diligence.

Public commitments are often the most successful. By their very nature they are harder to go back on. Whisper that you're planning to get up early and go for a run, and it's the easiest thing in the world to wake up, take a look at the chilly morning and burrow back under the covers. Make a pact with a friend to meet to run together, and, chilly or not, when morning comes around, you'll be lacing up your running shoes.

Scientists in Iowa found that public commitment was the most effective in an experiment to get householders to cut their gas usage.[78] They divided their householders into three groups, with each household receiving a 20-minute visit from a researcher, in which:

⊙ **Group 1** were given simple **tips to save energy** in the home
⊙ **Group 2** were asked for a **verbal commitment** to reduce energy consumption
⊙ **Group 3** were told that their names were being published in the local newspaper as examples of 'energy-conserving citizens' – a **public commitment**

After a month, meter readings revealed that while neither the first nor second group had made any significant reductions in consumption, the third group had cut usage by 12.2 per cent, on average. The very fact that they were going to have their names published in a local paper ensured the group's commitment to the project. Interestingly, even after they were told that their names would no longer be published, their energy-saving behaviour went from strength to strength in the following year, with an average reduction of 15.5 per cent. Once they started, they couldn't stop; habits had altered and their new behaviours had become ingrained. Something worth considering if you're embarking on a fitness regime, perhaps.

However, some commitment-bias approaches might not be to everyone's taste:

⊙ **Blackmail yourself:** A Facebook page by the Dutch anti-smoking council is designed to help you quit smoking. You set the date you want to quit, upload embarrassing photos of yourself to a special file and nominate a friend who has access to these and who will make them public if they ever catch you with a cigarette in your hands.
⊙ **Donate money to something you hate:** Imagine committing to a contract that will require you to donate money to a thoroughly horrible cause (e.g. the American Nazi Party) if you fail to deliver on your promise to yourself – it's a scarily powerful motivator.

Irrational Escalation Bias, also known as the Sunk Cost Fallacy OR How to throw good money after bad

This is the tendency to make irrational decisions based on what seemed to be rational decisions in the past. This is most likely to relate to financial matters, but could easily be emotional, effort-based or have involved a long-term commitment – it's where we find ourselves in situations where we feel we've come too far to pull out now, so we not only keep on doing what we're doing, but escalate our involvement because pulling out could imply our initial decision was misjudged and all that money, emotion, effort and time wasted. Whether it's holding on to shares that are crashing, continuing to gamble when we've nothing left to gamble with or sticking with a relationship which is well and truly over, irrational escalation bias is not our friend.

This bias was first described by Barry M. Staw at the University of California, in his 1976 paper 'Knee-deep in the Big Muddy: A Study of Escalating Commitment to a Chosen Course of Action'.[79] Staw sets the scene thus:

> It is commonly expected that individuals will reverse decisions or change behaviours which result in negative consequences. Yet within investment decision contexts, negative consequences may actually cause decision makers to increase the commitment of resources and undergo the risk of further negative consequences.

Staw's study identified the behavioural bias as 'a self-justification process in which individuals seek to rationalise their previous behaviour or psychologically defend themselves against adverse consequences'. He further identified that an individual committed the greatest amount of resources to a previously chosen course of action when they felt personally responsible for its negative consequences. In other words, the deeper you're in the sh*t when it's all your own fault, the greater the chance that you'll keep on digging. The phenomenon and the sentiment underlying it are reflected in such financially inspired idioms as 'throwing good money after bad' and 'in for a penny, in for a pound'.

More recently, the term sunk cost fallacy has been used to describe the phenomenon whereby people justify increased investment in a decision based on cumulative prior investment, despite new evidence suggesting that the cost, starting today, of continuing the decision will outweigh the expected benefit; we've put so much into something, we can't bear to acknowledge that it might be time to call it a day. Such investment may include money, time or – in the case of military strategy – human lives.

An example of irrational escalation (often used by US psychology professors to demonstrate the concept to first-year students) is the dollar auction experiment.[80] The set-up involves an auctioneer who volunteers to auction off a dollar bill with the following rule: the dollar goes to the highest bidder, who pays the amount they bid. Nothing unusual there you might say. The twist is that the second-highest bidder must also pay the highest amount they bid, but gets *nothing* in return.

> Suppose that the game begins with one of the players bidding 1 cent, hoping to make a 99-cent profit. He or she will quickly be outbid by another player bidding 2 cents, as a 98-cent profit is still desirable. Three cents, same thing. And so the bidding goes forward.
>
> As soon as the bidding reaches 99 cents, there's a problem. Remember they're bidding for a dollar bill. If one player bid 98 cents, he or she now has the choice of losing the 98 cents or bidding $1.00, for a profit of zero. Now the other player is faced with a choice of either losing 99 cents or bidding $1.01, and only losing one cent.

After this point, the two players continue to bid the value up well beyond the dollar, and neither stands to profit.

By the end of the game, though both players stand to lose money, they continue bidding the value up well beyond the point that the dollar difference between the winner's and loser's loss is negligible; they are fuelled to bid further by their past investment.

We can find it hard to admit we have made a mistake, in any context. The real question is, how bad must our losses be before we change course, and does knowing about irrational escalation bias empower us to do anything about it? Say you buy a lottery ticket in the hope – your brain is in thrall to the optimism bias, after all – of winning. Perhaps you've been doing it for years (in fact, you could have been doing it since 19 November 1994, when the first draw took place), and you may have won absolutely nothing in all those years. What keeps you buying the tickets? In your mind, you've invested too much to give up now, and it must soon be your turn to win – mustn't it?

Negativity Bias: Why we can't get negative stuff, however small, out of our minds

Negativity bias is the phenomenon that ensures we pay more attention, and give more weight, to negative experiences (e.g. thoughts, emotions, events) than positive experiences or information. It's not that we're all mad for gossip or chock-full of schadenfreude (not completely, anyway); it's psychological, too. Our brains have a greater sensitivity to unpleasant news.

This is why we tend to think about the bad stuff for far longer than we do about the good. It's why insults hurled at us years ago still shine bright. It's why we hold grudges. (A friend in her fifties can still recall with sharp clarity the full name of the 'best friend' who called her 'Fatty' over forty years ago.) It's the reason political smear campaigns are horribly more effective than positive ones. Nastiness and negativity mess with our brains. In a 2001 article entitled 'Bad is Stronger Than Good', in which University of Queensland psychologist Roy F. Baumeister and his co-authors analyse the power of negativity, they make the point that over and over, in all sorts of circumstances, bad things make a bigger impact than good. Of course, one of the main reasons for this is evolutionary:

> A person who ignores the possibility of a positive outcome may later experience significant regret at having missed an opportunity for pleasure or advancement, but nothing directly terrible is likely to result. In contrast, a person who ignores danger … even once, may end up maimed or dead. Survival requires urgent attention to possible bad outcomes, but is less urgent with regard to good ones.[81]

Our capacity to weigh negative input so heavily most likely evolved to keep us out of harm's way. From the dawn of human history, our very survival depended on our skill at dodging danger. The brain developed systems that would make it impossible for us not to notice danger and, thus, be better able to respond to it fast. In today's world, we don't need to dodge physical dangers as regularly as our ancient ancestors did, but our negative reflex remains finely tuned nonetheless. Once we are conscious of its influence upon us, we can see it in action and perhaps tune it out when it pops up unnecessarily enthusiastically.

Baumeister and co. make the point that losing money, being ditched by friends or being criticised will undoubtedly have more of an impact on us than finding money, making new friends or receiving congratulations for a job well done. If someone says something critical or unpleasant to us, it can completely ruin our day, no matter how much good stuff preceded it or came after. It's always the nasty that bites deeper. Gretchen Rubin, author of *The Happiness Project* and *Forty Ways to Look at Winston Churchill*, blogged on this very point:

> I've discovered that reading one bad comment will ruin my morning, and reading five positive comments won't cheer me up. So, I try to resist.
>
> For example, within marriage, it takes at least five good acts to repair the damage of one critical or destructive act. With money, the pain of losing a certain sum is greater than the pleasure of gaining that sum. I know this from my own experience. I remember, for example, that hitting the bestseller list with *Forty Ways to Look at Winston Churchill* thrilled me less than a bad review of that book upset me.
>
> Research shows that one consequence of the negativity bias is that when people's thoughts are wandering, unoccupied, people tend to begin to brood; the negativity bias means that anxious or angry thoughts capture our attention more effectively than happier thoughts.[82]

Studies conducted at Ohio State University by John Cacioppo (who founded the University of Chicago Center for Cognitive and Social Neuroscience) demonstrated that the brain reacts more strongly to stimuli it deems negative. There is a greater surge in electrical activity. Thus, our attitudes are more heavily influenced by bad news than good. Think about the last time you lost some money – a £20 note, for instance. You're certain to have been more emotional about that, and it would have niggled you for longer, than if you unexpectedly came across the same amount in an old coat pocket.

Extrovert and larger-than-life twentieth-century socialite Bubbles Rothermere was reputed to have a sofa cushion embroidered with the sentence '*If you haven't anything good to say about anyone, come over here and sit by me.'* I guess we can all recognise the lure of a good bad story. How we love conversations that begin, 'I'll tell you another terrible thing that happened...' and 'Did you hear about so-and-so?'

If it's of any comfort, the older we get, the more we journey away from emphasising negative things as we focus increasingly on the positive, paying more attention to things that we can remember with pleasure. The scientific explanation for this is that the older we get, the better able we are to regulate our emotions and, since we know we have less life ahead of us, it is in our interest to make what life we have as positive as possible. And we can change things to make this happen. We might be able to change our environment to achieve a more positive outlook, and we can certainly amend our expectations. Psychologists have found that a healthy and positive approach to ageing can, in effect, be self-stimulated, and their advice is to concentrate on short- rather than long-term goals, and to focus on the positive.[83]

Confirmation Bias: Aha! Just as I thought!

This is when we seek or interpret evidence in ways that support our existing beliefs.

Confirmation bias influences how we seek out information and research things; it makes us more receptive to, and more likely to take note of – and consider to be relevant and right – data that supports our own beliefs, while at the same time rejecting or simply overlooking conflicting evidence. It's confirmation bias, too, that nudges us to seek out positive, rather than negative, information to apply to our reasoning. It biases us against interrogating our hypotheses to test whether or not they stand up. And it's a bias that comes to the fore in politics; from Brexit to Trump, you don't have to dig very deep to find it hard at work. And it is *very* hard at work on social media. Of course, we enable it by following like-minded thinkers on Twitter and other social media platforms, curating our very own echo chambers.

A friend told me recently that she had once been a secretary of a local council committee that held monthly meetings. The meetings were taped but she took her own notes – comprehensive notes, she thought. In fact, she'd have gone as far as to say she always transcribed the meetings in full. That is, until she listened to the tape of one meeting to check something while looking back through her own notes. She was horrified to realise that the only parts of the meeting she had really covered 'in full' were the parts where she had agreed with what was being said.

Confirmation bias ensures that we view the world through a filter and practise very selective thinking. Most of us instinctively avoid evidence that contradicts our opinions. Contrary information is upsetting and confusing. We don't want to admit our beliefs may be wrong and it can be inconvenient to have to form a completely new point of view.

> What the human being is best at doing is interpreting all new information so that their prior conclusions remain intact.
> Warren Buffett, billionaire US business magnate

The problem with confirmation bias is that we are potentially closing ourselves off to a whole raft of new ideas and to new angles on old ones.

Let's go to Karl Popper, the great philosopher of science, for two useful and illuminating angles on confirmation bias:

1. It is easy to obtain confirmations, or verifications, for nearly every theory – if we look for confirmations. Confirmations should count only if they are the result of risky predictions ... A theory which is not refutable by any conceivable event is non-scientific. Irrefutability is not a virtue of a theory (as people often think) but a vice. Every genuine test of a theory is an attempt to falsify it or refute it.[84]

2. Great scientists ... are men of bold ideas, but highly critical of their own ideas; they try to find whether their ideas are right by trying first to find whether they are not perhaps wrong. They work with bold conjectures and severe attempts at refuting their own conjectures.[85]

You might want to bear this in mind next time you find yourself leaping to an apparently inevitable conclusion. Before asking yourself to confirm why you're right, try looking for reasons why you may not necessarily be quite as right as you think you are; why, in fact, you might be wrong. This could be applicable to discussions with teenage children about what's acceptable and what's not, or in team meetings to ensure that all perspectives are given equal weighting.

The *New York Times* tackles the issue head-on with a practice of directing subscribers to read contradictory views. Here's op-ed columnist David Leonhardt:

> I've encouraged readers before to spend time reading arguments with which they disagree. So if you were disappointed by Brexit (as I was), be sure to read, 'The Good News on Brexit They're Not Telling You,' by Daniel Hannan, whom the *Financial Times* has called 'The brains behind Brexit'.[86]

Leonhardt goes further, suggesting that if we don't consider an argument from all sides before discussing it, we simply convince ourselves of our own rightness: 'The more we talk politics, the more confident we can become that we're right.' His remedy fits nicely here: 'Pick an issue that you find complicated, and grapple with it. Choose one on which you're legitimately torn or harbour secret doubts. Read up on it. Don't rush to explain away inconvenient evidence.'

In a study conducted in 1983, seventy Princeton undergraduates were asked to assess the academic ability of a ten-year-old girl.[87] First, half the undergraduates watched a video that indicated that the girl was from a high socio-economic background, while the other half watched one that indicated she was from a low socio-economic background. Next, both groups watched a further video of the girl responding to twenty-five questions on maths, reading and science. Participants were then asked to assess the girl on different achievement categories (reading comprehension, language skills, and so on). Participants rated her performance significantly higher when they thought she was from a higher socio-economic background. They had formed a hypothesis about her abilities assuming there to be a relationship between socio-economic status and academic ability; they then interpreted what they saw in the questions video in ways that confirmed their hypothesis.

Depending on where you stand in the Brexit debate, you'll find yourself more (or less) in agreement with a *Sunday Times* article by the paper's science editor, Jonathan Leake, with the headline 'Science Has Dim View of Brexit Voters' Brains'.[88] High-profile Brexiteer Nigel Farage responded by describing the research as 'divisive and arrogant. Remain voters may have higher IQs but I'm not sure many could boil an egg or set up a business.'

But then he *would* say that, wouldn't he?

Loss Aversion: Why losing something hurts more than a similar gain

We feel the pain of loss more than we feel the elation of gain. It means that most people are motivated more by avoiding a loss than acquiring an equivalent gain.

We've already commented on the fact that we would be much more upset to lose £20 than we would be to find the same amount in an old coat pocket. This is partly due to our tendency to focus on the negative, yes, but it is also because of loss aversion, whereby we try to avoid losing where we possibly can. In effect, it's the *not losing* that motivates us more than the winning.

It's a bias that helps to explain why investors are often slow to sell shares that have lost value. Likewise, it might be a factor for explaining why companies have trouble disposing of bad business assets. Richard Thaler and Cass Sunstein had this to say about it in their book *Nudge*: 'People hate losses. Roughly speaking, losing something makes you twice as miserable as gaining the same thing makes you happy … Consequently, loss aversion produces inertia, meaning a strong desire to stick with your current holdings.'[89]

The concept of loss aversion also means that losses loom far larger than gains in our minds, and this can create irrational behaviour. Let's look at how this can work in real life. Think about buying something on eBay; you're in a bidding war and the seconds are ticking away. Another bidder keeps increasing the bid; you're losing, so you keep upping your bid to overtake the other bidder, and suddenly it's less about wanting to win the item, and more about not wanting to lose it.

Loss aversion explains why studies show that paying by cash hurts more than paying by credit or debit card, and may be a useful insight for those whose job it is to encourage us to use our credit cards more often. When we pay for something with cash, the pleasure of the purchase is slightly undercut by the pain of giving away the hard currency; we physically experience the loss of our money. When we pay with credit and debit cards, we spend the same funds but via an invisible transaction that anaesthetises the pain. If you're trying to cut down on spending, use cash, not cards.

One of the main areas where loss aversion works hardest is in the stock market. You've invested in specific shares for years, but they're not performing well, the share price is dropping, you're losing money as each day passes. What should you do? What do you *feel* you should do? If you sell, you'll take a hit. Surely the shares will soon pick up? You've invested in them for so long…

A study conducted by Terrance Odean, a professor at the Haas School of Business at the University of California Berkeley, analysed investment data on 10,000 investment accounts managed by a US brokerage firm over the period 1987–93.[90] Odean found that investors suffer from loss aversion and generally hold on to losing stocks for longer than winning stocks – they held on to losing stocks for a median of 124 days, but held on to winning stocks for a median of 102 days. The rationale was that the losing stocks would 'bounce back'. They also sold more winning stocks than losing stocks, selling an average of 15 per cent of winning shares, but only 9 per cent of losing shares. Odean analysed to what extent the decisions made to sell or to stick made financial sense. He found that the sold stock outperformed the unsold stock by an average of 3.4 per cent per annum.

If you're a professional golfer, an ability to drive your loss aversion into the long grass could reap financial rewards, as revealed in a study conducted by researchers Devin Pope and Michael Schweitzer.[91] They drew on extremely rich data on the play of professional golfers collected by the PGA Tour. The PGA Tour hires 250 people to gather detailed information about play, using lasers around each hole to measure and record – to within an inch – the coordinates of each ball before and after every shot. Pope and Schweitzer used this data set, analysing more than 2.5 million putts, focusing specifically on birdie putts (one shot less than par) and putts to make par (the typical number of shots a professional would make to get the ball in the hole). They found that players focus more on the par shots than the one under-par shots, to avoid 'dropping a shot', and make birdie shots 2 per cent less often than par shots. 'Gaining' a shot (by making a birdie putt) is of less importance to them in a tournament. Tiger Woods explains it like this:

Any time you make big par putts, I think it's more important to make those than birdie putts. You don't ever want to drop a shot. The psychological difference between dropping a shot and making a birdie, I just think it's bigger to make a par putt.[92]

So, what does a difference of 2 per cent translate to? It doesn't seem as if it would make a huge impact, does it? The researchers tracked down PGA Tour statistics and estimated that if the top twenty professional golfers were to be unaffected by this type of loss aversion and make birdie putts as often as they made par putts, their tournament score would improve by more than one stroke per tournament. This is, in fact, a substantial difference – on average, these golfers would have earned an additional $640,000 per year from PGA earnings, not including further income from endorsements and commercial sponsorship.

Endowment Bias or Endowment Effect: How ownership creates satisfaction

This is where people place a higher value on an item they own than on an identical item they do not own.

The power of the endowment effect was famously quantified in an experiment conducted by psychologists Daniel Kahneman, Jack Knetsch and Richard Thaler.[93] In the experiment, Cornell University students were randomly chosen to receive mugs (which sold for $6 each in the university shop). Those with mugs were asked to name the minimum price at which they would sell their mugs; those without were asked to name the maximum price at which they would buy the mugs. Students with mugs offered them for sale at about $4.50; students without mugs offered to buy at about $2.25. The median selling price was more than twice the median offer price. Ownership of the mugs increased their perceived value.

The problem with the endowment effect is that it doesn't apply only to things. It can apply to our opinions too. As Dan Ariely, a professor of psychology and behavioural economics at Duke University, North Carolina, puts it:

> Ownership is not limited to material things. It can also apply to points of view. Once we take ownership of an idea – whether it's about politics or sports – what do we do? We love it perhaps more than we should. We prize it more than it is worth. And most frequently, we have trouble letting go of it because we can't stand the idea of its loss. What are we left with then? An ideology – rigid and unyielding.[94]

Both loss aversion and the endowment effect can stall us. We don't sell poor-performing shares (hey, they might bounce back), we hang on to old points of view, we overvalue the stuff we have, and the resulting stagnation can get in the way of new experiences and new understanding.

Availability Bias OR *Why we should be more afraid of putting on our socks*

Availability bias is about our tendency to estimate what is likely to occur based on what is more available in our memory. The stuff we have uppermost in our memories is biased towards the vivid, the unusual, the emotionally charged moments and events and, very often, the bad stuff. And if something is vivid and easy to recall, it seems most likely to happen.

In daily life this means we'll forget what we had for breakfast every day last week, but we'll remember that there was a car accident right outside our house on a Tuesday three months ago. We have breakfast every day, but the rare event of the car accident will be foremost in our minds for a while. This memory bias makes us anticipate the likelihood of something happening, or being likely to happen, based on how easily an example can be brought to mind; essentially, the vividness of the memory makes it stick with us. Terrorist attacks are a contemporary benchmark for this. Attacks on people travelling on public transport or at pop concerts are horrific and availability bias prompts us to anticipate repeated attacks in the same context. We fear the possibility of these events, but we should probably be more concerned about things closer to home. Our clothing represents a much more credible threat to our well-being, with around 10,000 people each year in the UK facing a stay in hospital after an incident caused by putting on their socks. A further 6,000 are taken to hospital after a trouser-related catastrophe, while over 3,000 individuals have an unhappy experience with a clothes basket.[95] But these domestic risks worry us not a jot in comparison with the fear of terrorist threats, tornadoes and tsunamis, though each one poses far less of a risk than the socks, trousers or clothes baskets.

And yet again it seems the availability bias is an evolutionary trick. It's hardwired into us and has been for aeons. Here's David Myers, a professor of psychology at Hope College, Michigan:

> Human emotions were road tested in the Stone Age. Yesterday's risks prepare us to fear snakes, lizards, and spiders, although all three combined now kill only a dozen Americans a year. Flying may be far safer than biking, but our biological past predisposes us to fear confinement and heights, and therefore flying.[96]

News organisations seek out sensational news, and this fuels our availability bias and makes us think that these stories are more normal and representative of the bigger picture than is really the case. Journalist David McRaney, who started out as a newspaper reporter covering Hurricane Katrina and wrote the international bestseller *You Are Not So Smart*, explains it thus:

If you see lots of shark attacks in the news, you think, 'Gosh, sharks are out of control.' What you should think is, 'Gosh, the news loves to cover shark attacks.' The availability heuristic shows you make decisions and think thoughts based on the information you have at hand, while ignoring all the other information that might be out there.[97]

For Hillary Rodham Clinton, it was most likely availability bias that put paid to her chance of the US presidency in 2016:

Imagine you're a kid sitting in history class thirty years from now learning about the 2016 presidential election which brought to power the least experienced, least knowledgeable, least competent President our country has ever had. Something must have gone horribly wrong, you think. Then you hear that one issue has dominated press coverage and public debate in that race more than any other. 'Climate change?' you ask. 'Health care?' 'No,' your teacher responds. 'Emails.'[98]

There was, it seemed, no way of closing 'those damn emails' down. The more they were discussed in the media, or explained or apologised for by Clinton and her team, the more resolutely they became fixed in people's minds as indicative of some sort of shady behaviour. The more we hear about something, the more we assume it to be true.

In lots of different contexts we don't think in terms of statistics, we think of memorable examples (both desirable and not so desirable) and we overestimate the likelihood of these things happening to us, too; so, we imagine buying lottery tickets and winning because we've read the stories about lottery-ticket-winners who have scooped the jackpot, and we're probably nervous about our imminent long-haul flight because when we hear about planes in the news it's about planes crashing and passengers dying, and we pace the floor because our teenage daughter isn't home and it's late and all the papers are full of late-night attacks on young women. In each of those cases it's the availability heuristic that's behind it all, making us worry irrationally. If you need to find more solid thinking ground, try to dismiss your instinctive response to an idea and ask yourself: what's the more likely outcome, based on probability?

Probability will likely prompt us to have serious concerns about the impact of climate change on our security. It's a high-level known risk and science backs that up, but it's a low-level dread risk: we are not fearful of being in immediate peril because of it. Terrorism ranks high on the dread risk table but is a low-level known risk. We're much more afraid of being killed in a terrorist attack (unlikely) than being killed by climate change (increasingly likely).

Scarcity Bias: What's that? It's the last one? Then I must have it!

This is based around the economic principle that scarce resources are more desirable.

It might seem pretty standard stuff but it's also immensely powerful. If we know or think there isn't much of something, whatever the thing, it's suddenly much, much more desirable. We want what we can't have – it's as simple as that. That's why the words 'limited edition', 'rare' and 'one of a kind' have such cachet and it's why 'only one left' (see Amazon or booking.com) can prompt a quick purchase. I nearly bought a terracotta pot in a garden centre the other day simply because an assistant I had asked about the pot called out to a colleague, 'This Vienna pot... It's the last one, isn't it?' And recently I went online, and on the phone at the same time, to 'queue' to buy some items from a limited-edition range of designer clothes that were being launched that day by a high-street store. I waited in my virtual line for twenty minutes before getting through and then proceeded to buy more than I had intended to – *just in case* it all sold out. It was limited edition – what other choice did I have? I also bought a book precisely because Amazon 'told' me there was only one left. After I bought it, I left the site and refreshed my page – and when I went back to check, there was still 'only 1 left in stock'. Go figure. It's illogical but we are easily swayed by it. Psychologists call it the anticipation of regret – and it's hard to resist its thrall.

When online clothing retailers list new stuff, they'll likely display all available sizes along with the sizes that have completely sold out. It's a way of hurrying you along to purchase ('This thing is running out') and reassuring you at the same time ('It's popular – look how stocks are falling).

The online fashion brand Supreme plays to this bias by scheduling a weekly 'drop' of newly listed items at precisely 11 a.m. UK-time each Thursday. Customers are allowed to purchase one item in a particular line and one only – and it's a race against time to secure it. Customers know this; it's part of the purchase deal. Too slow and you lose out. If you're quick enough you can wallow in the smug pleasure of seeing a big fat 'SOLD OUT' plastered across the image of the merchandise you bought.

Scarcity bias encompasses a number of factors: there's the fear of missing out (FOMO: loss aversion), the pressure of other people's actions (stuff is scarce because other people think it's good and have got there first – a kind of social norm) and our sense that if we don't push on and buy or book it right now, we'll regret it. In anticipating the sorrow of NOT buying or booking, we nudge ourselves into doing it.

But – and it is of course a big but – scarcity bias prompts us to act fast, without necessarily carrying out the 'due diligence' that some decisions or purchases might require. When you find yourself worrying about missing out, take a breath: you're probably just being pressured into it.

The Peak-End Rule: Why the last moment in an experience really does matter

The peak-end rule – first suggested by Daniel Kahneman and colleagues[99] – is a memory heuristic where we judge our past experiences almost entirely on how they were at their *peak* (the most intense moment of the whole thing – good or bad) and how they *ended*. This fits into the bias portfolio because we overweigh the impact of these moments and then apply this to the experience as a whole. It can apply to both positive and negative experiences.

Whenever you host a social occasion, bear in mind that it will be evaluated or remembered by your guests as a kind of average of the *most impactful* bit of the occasion and the *end* of the occasion, resulting in an overall evaluation that falls somewhere in between those two points.

There's a pretty key factor in the peak-end rule: we apply it retrospectively. Therefore, it's the *remembering self* who rates the event, *not* the *experiencing self*. Thus, as Daniel Kahneman points out, the remembering self can rob an event of all satisfaction by focusing on an unsatisfactory ending:

> A comment I heard from a member of the audience after a lecture illustrates the difficulty of distinguishing memories from experiences. He told of listening raptly to a long symphony on a disc that was scratched near the end, producing a shocking sound, and he reported that the bad ending 'ruined the whole experience'. But the experience was not actually ruined, only the memory of it. The experiencing self had had an experience that was almost entirely good, and the bad end could not undo it because it had already happened.[100]

Ryanair celebrates each flight that arrives on schedule with a little trumpeted announcement. It's jolly and it feeds off passengers' relief at having landed in one piece. Their peak-end memory of the flight? Smiles all round. Similarly, Qantas recently swapped full inflight meals for a light snack, but the air stewards completed the service by handing out Lindt chocolates. Another peak-end.

If you want to host an unforgettable party, make sure you engineer a big finish. With any party gathering, you can do a lot to raise remembered enjoyment levels by enhancing the end of the event. Remember: it's the peak-end rule. Think about the end of your event and tailor a little special something if you want to be sure of success.

Say you had a pretty run-of-the-mill dinner party – good food, relaxing company but nothing particularly out of the ordinary; you could greatly enhance your guests' memories of the evening by sending them all home with a goody bag of some kind – unexpected, pleasurable, a memorable endpoint that boosts the overall evening-evaluation. After all, what do all young children discuss and treasure after a party? Is it what was in the sandwiches? Who won the games? How successful the entertainer was? No – they talk about what they got in the party bag. The big finish.

In the same way, after the same kind of evening, the end of the occasion could be overwhelmed by a massive argument triggered by you or one of your guests – an entirely unexpected, alarming and memorable endpoint that downgrades the overall 'enjoyment'-evaluation at the time (although it boosts the gossip potential for the next day and, with the benefit of retrospective impact bias, becomes increasingly enjoyable as time goes by).

You can apply the peak-end principle to gift-giving, too. Best advice? Give one great gift. Don't give a great gift and something extra, because the something extra will downgrade the great gift. Think about it in mathematical terms: the average of two gifts is $(x + y) \div 2$, whereas the pleasure of one gift (a great one, remember) is indivisible. If you *do* have two gifts to give, just make sure you give the best one last!

So, try working a big finish, or at least a slightly bigger one, into simple things you do every day, even something as simple as how you sign off from an email or say goodbye. Who knows what it might start?

The Bias Blind Spot: I have my faults, but being biased isn't one of them

We've now had a run-through of some of the most influential biases, and you can probably begin to see them at work on your own brain... Or can you? Because there's also a bias called the bias blind spot,* so called because it means we often don't see the impact of biases on ourselves – on our attitudes and behaviour.

The sneaky thing about the bias blind spot is that, while it ensures we remain confident that our judgements aren't biased (*Biased? Us?*), it manages to leave us free to point out the biases that prompt the behaviour of others. For example, you're watching a hockey match and your son or daughter is playing on one of the teams. The match referee is the opposing team's coach. If your child's team starts to lose the match, how likely is it that you will feel that the referee's decisions are biased in favour of their own side? But obviously we don't see that as *our* bias (i.e. our dismay that *our* child's team is losing) – it's the ref who's biased, surely? Be honest: for each of the biases covered so far in this chapter, how many times have you heard a little internal voice saying, 'Well, that doesn't apply to me...'?

Emily Pronin, an associate professor of psychology at Princeton University, and Matthew Kugler, an associate professor of law at Northwestern University who is also a psychologist, suggest this gap between our lack of ability to perceive our own biases, and the sharp focus in which we hold the (blatant) biases exhibited by others, lies in our ability to objectify our own responses, regarding them as free from bias, based on our introspective assessment that we always try to be fair in our judgements.[101] We know, too,

---

* The term 'bias blind spot' was created by Emily Pronin, Daniel Lin and Lee Ross, psychologists at Princeton University.

that being *un*biased is the better thing to be, so that's what we believe ourselves to be. However, since we are not privy to the introspections of others, and neither are we likely to value them as we do our own, we see their behaviour as comparatively subjective – in other words, we don't think other people are as nice as we are *or* as right-thinking, so we don't credit them with the desire to be unbiased.* As Pronin and Kugler point out.

> When our colleague judges the average work of his friend as better than the stellar accomplishments of a mere acquaintance, or when our neighbor argues that the new bright-red fire hydrant should go in front of any house but hers, we are struck by those individuals' blindness to their own biases.[102]

None of us is immune from the influence of bias. Daniel Kahneman, the father of behavioural science, said as much to his audience of world leaders at the World Economic Forum at Davos in 2013, where on the opening morning he talked about the science behind human decision-making and cognitive biases, outlining some of the powerful ideas, theories and concepts developed with his colleague Amos Tversky.

But Kahneman said that, after more than forty years of studying the subject, even he is not immune to bias: 'My intuitive thinking is just as prone to overconfidence, extreme predictions and the planning fallacy as it was before I made a study of these issues.' He has also remarked on the fact that many of his academic peers (particularly the economists among them) regularly fall for the very same biases.

In a study by Emily Pronin and colleagues at Stanford University called 'The Bias Blind Spot: Perceptions of Bias in Self Versus Others', the findings pointed to the danger of the bias

---

* Note the influence of the egocentric/self-serving bias here too.

blind spot.[103] In a political context the bias blind spot can lead us headlong into conflict. Take the aftermath of the Brexit referendum, which resulted in family feuds, each side equally convinced of their own rightness and of the wrongness of those who voted differently. Pronin and co. explain exactly how this can happen:

> We find that our adversaries, and at times even our peers, see events and issues through the distorting prism of their political ideology, their particular individual or group history and interests, and their desire to see themselves in a positive light. When we reflect on our own views of the world, however, we generally detect little evidence of such bias. We have the impression that we see issues and events objectively, as they are in 'reality'.[104]

As long as we hold fast to the misguided idea that it is *other* people's attitudes that are misguided, while ours are steadfastly on the straight and narrow, conflict will result. Pronin and Kugler suggest we should work harder to acknowledge that we are all labouring under the same bias blindness. Indeed, when we criticise others for NIMBYism and then reject a proposal to build sheltered housing in our neighbourhood, we are wading knee-deep in the bias-blind-spot sea.

Still thinking, 'Well, I'm not like that. This doesn't affect me'? Rest assured, your bias blind spot is hard at work.

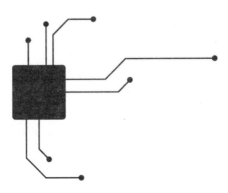

Section 2: How companies and governments are playing to our brain wirings to influence how we think and behave

In this section we will lift the cloak of invisibility to reveal how our behaviour is being subtly influenced by companies and governments 24/7 and to see how the four foundation stones provide a powerful toolkit for organisations to nudge and steer our behaviour. They're doing it all around you, and on a daily basis, too: on your social media, on your phone, in your car, in the supermarket, in the direct mail you receive, in your online life, in advertising. Don't be alarmed – it's not necessarily a bad thing, but it's good to be able to recognise it when it's happening.

Governments typically rely on biases to influence us. For example:

⊙ Social norms/herd instinct – they like to suggest that if other people are doing something, we should be doing it too.

⊙ Egocentric bias – making it personal can deliver a more powerful message.

⊙ Status quo bias/default settings – they'll leave the desired behaviour as the default option, ensuring that choosing to go against the default requires us to take deliberate action, something they anticipate lots of us won't do.*

Commercially oriented companies often make use of a broader palette of options, such as:
⊙ Reciprocity bias
⊙ Peak-end rule
⊙ Scarcity bias
⊙ Loss aversion
⊙ Social norms
⊙ Status quo bias/default settings
⊙ Anchoring

The adoption and application of behavioural science is set only to grow, and many new jobs in the field will be created. There are already numerous organisations advertising behavioural-science focused jobs based in a variety of industries – from social media and finance to pet rescue and nature conservation.

Let's look first at how commercial companies are harnessing these brain-wiring insights to nudge and steer our behaviour.

Making us well disposed towards them

Companies want us to like them. If we like them, we'll buy stuff from them. Some of them leverage our biases to make this happen. When we receive a gift from someone, we think nice things about them, it makes us warm to them and more likely to deal positively and happily with them in the future. From personal experience, it would seem that Marks & Spencer frequently apply this technique

---

* A few years ago, Finnish cybersecurity firm F-Secure tested the likelihood that customers ever read the Terms and Conditions by inserting a clause into their T&Cs attached to an offer for free Wi-Fi in London. The T&Cs stipulated users would have to hand over their eldest child 'for the duration of eternity'. Six people signed up before the page was disabled.

during the checkout process in their stores. Next time you're packing your shopping, wait and see if the cashier makes a positive comment (their 'gift' to you) about something you've bought: 'What delicious looking strawberries!' perhaps, or 'Aah, how sweet...' for the chocolate teddy bears. (One member of staff in my local M&S store has adopted an all-encompassing 'Enjoy your lovely things!' as he bids his customers farewell.) I believe they've been trained to say one nice thing per load of shopping on the conveyor belt, just as they have been trained to say 'Thank you for waiting' (even if you haven't waited very long at all) and to offer to help us with our packing. It makes us not complain. We smile. We say, 'No problem.' We feel liked and we like them back in return. If we think about it, our experience with the store has been run in parallel with the peak-end rule bias (which means we will judge our experience based on an average of the best bit and the end) with a little bit of reciprocity bias thrown in – they've been nice to us, so we feel kindly disposed back, and the ripple of positivity spreads.

Playing the FOMO card

Of course, they don't just want us to *like* them; they want us to buy things from them too, and they want us to do it *now* rather than thinking about it and then not doing it. They want us to be prey to the Fear of Missing Out. So, they need to apply degrees of manipulation to get us to play ball.

Companies using behavioural nudging are particularly active online. Here's a quick look at my morning's email inbox and the subject lines from commercial companies:

⊙ *Time is ticking. Don't miss out!*
⊙ *Last chance! Sale ends tonight.*
⊙ *FLASH discount. 10% OFF ends 8 a.m.*
⊙ *Last day to play. How much will you save?*
⊙ *Snap me up before I'm gone!*

If we check something out just *once* online we'll be plagued by insistent reminders of that thing each time we go to answer an email or to visit Twitter or Facebook, as images of the shoes we briefly looked at, or the coat or the holiday, repeatedly pop up in the margins of our sight-lines. The online world exists to service our needs and desires, and it does it relentlessly, drawing on behavioural biases *all the time*.

The online retailer I Want One of Those.com sent me an email with the subject headline '*Hurry, 30% off everything for first 80 customers. Plus, free UK home delivery!*' I was attracted by the prospect of being one of the 'first 80' (they'd used scarcity bias on me – 'Help! If I don't act fast I'll lose out!'); plus, there's a huge appeal in the chance of being one of eighty people singled out for special treatment, so it appealed to me in a first-past-the-winning-post kind of way. The whole thing is a perfect case of loss aversion – 'If I'm not one of the first eighty, I won't just lose my 30 per cent off, I'll have to pay for P&P too.' Note how at this stage I don't envisage not buying *anything* – they have me already imagining the money I'm about to save or about to risk losing. In fact, they had me at 'Hurry'! So, I went on to the site and bought something. At the checkout, I put in my promotional code and, lo and behold, I 'qualified' for the 30 per cent discount. So, I bought something else as well. My behaviour had just been well and truly nudged and steered.

Bamboo clothing company BAM have a rather charming inducement (NB it's reciprocity bias) to get you to purchase. Here's what happened with me. I bought some socks and other stuff in the sale. I bought the socks because a few months ago they sent me a free pair of socks, out of the blue, no strings. Well, I say 'no strings' because at the time I didn't notice the barely visible reciprocity-bias-nudging strings attached to the 'free' socks. The socks were very comfortable. I decided to get another pair (or four) and that other stuff (in the sale). All good – job done. Now here's the charming bit. I got an email from them with the subject header '*Forgotten anything?*' followed up by this message:

We hope you received everything from your order in good time and to your satisfaction. Do you often find that once you've ordered you think, I wish I added this or that, but don't want to pay postage again? No problem, as this offer code will enable you to do just that. So, if you are interested use the offer code: aw-8919rq...

Happy shopping from everyone at BAM Bamboo Clothing.

I was so impressed by the thoughtful approach (and, let's face it, postage *is* a pain), I went back and took advantage of that offer code. It's reciprocity bias working discreetly, politely and well.

Next time you go to the Amazon website, make a mental note of the number of social norm-, scarcity bias- and loss aversion-based nudges there are in operation. For example:

- *Only 1 left in stock – order soon* – this definitely works! This is scarcity bias and loss aversion rolled into one.
- *Customers who bought this item also bought* – this works on our social norms/herd instinct: we feel safer if other people thought these items were a good thing to buy.
- *Frequently bought together* – again, this is our social norms/herd instinct at play, saying the same thing in a slightly different way.
- *Get it by tomorrow if you order in the next 1 hour* – a form of loss aversion: if you're too late, you'll have to wait until the next day.
- *5-star customer reviews* – allows us to feel that we are choosing wisely and advisedly: if others think it's OK, that's good enough for us, even though we have no idea at all about the other people whose recommendations we rely on.

Booking.com is a good example of the way FOMO-based biases – the most obvious being loss aversion or scarcity bias – can be used in marketing, with tags like:

⊙ *Last chance! Only 1 room left.*

⊙ *Most recent booking for this hotel was fourteen minutes ago.*

⊙ *This hotel is likely to sell out very soon!*

⊙ *There are 11 people looking at this hotel!*

And the site also lists rooms that have sold out, rather than not listing them at all. These messages all do their job of raising our anxiety levels as we search the site, making us fret that the hotels are so popular that they could sell out before our very eyes – literally *worrying* us into making a booking faster than we had intended. And they use other biases, like anchoring (showing us the RRP and the discounted price), social norms (telling us how many bookings they've already had today) and discounting the future (letting us book now and pay later).

Like Amazon, they also harness the social norm bias, not only by including reviews of hotels by other users, but, crucially, by telling us which country other users are from. This appeals to in-group biases, as we are more likely to trust people whom we feel to be 'like us'. Plus, we want to be part of something; we want to be just like other people. We want to be part of the herd.

I recently had a scarcity bias prompt from a sportswear retailer. I'd put some items into my online shopping basket on their site, prompted by a code for 15 per cent off that they had emailed. But I didn't complete the purchase. The clever retailer *knew* I had stuff in my basket that I hadn't gone ahead and bought yet, so they emailed me to say: '*You've left some items in your shopping bag. Hurry back before stock runs out.*' See what they did there? They tried to get me on loss aversion and scarcity bias. If I check on my emails from retailers, they all seem to be communicating in the same way: 'Hurry up, or *you might miss out.*'

Here's Crabtree & Evelyn with a recent weekend offer covering loss aversion and scarcity bias (again):

Snooze you lose. 20% off everything! Don't miss out. Shop from home this weekend and you'll receive 20% off your entire order. Once it's gone... it's gone.

Online ticket seller Viagogo is a manic user of social norm and scarcity bias prompts, nudging (with a capital N) customers to buy tickets.* All of the following messages popped up in one purchase experience as reported in an article in *The Times* in 2018:[1] *Selling Fast*; *Secure Your Seats*; *Don't Miss Out*; *Prices Rising*; *Many People Viewing*; *Fast Checkout*; *Wide Selection*. Viagogo, as it happens, is one of the ticketing sites that adds in substantial extra costs to the price of a ticket – this is known as drip pricing. The customer who described her frenetic and unhappy purchasing experience with the company subsequently discovered she had paid more than £75 for a ticket with a face value of £30. She commented:

> The price was shown in a small box on the left of the screen, while a large box flashed on the main screen saying I only had three minutes to finish the transaction. The initial screen I looked at online said that only 1% of tickets were left, a blatant lie because the same claim was being made several days later.

The learning from this: take a breath before panic-clicking.

You'll have noted that most online marketing tactics are FOMO-related – a combination of scarcity bias and loss aversion right there. Not many left? FOMO. Other people are already all over this, plus there aren't many left? FOMO. This offer won't last for ever, plus other people are already all over this, *plus* there aren't many left? FOMO. FOMO. FOMO. And some marketers

---

*   It was also on the receiving end of an enforcement order from the Competitions and Markets Authority in 2018 and has been told to tighten up its business practices to increase customer protection measures.

take FOMO to a whole new level of exclusivity, creating VIP clubs or invitation-only previews. Not signed up? Not invited? You're definitely missing out!

Directing us to the 'safety' of the herd

Social norms are incredibly powerful. For instance, imagine that you're looking for somewhere to eat lunch. You come across two restaurants side by side. One is buzzing, with lots of people and only a few free tables, while the other is completely empty. Which one would you choose? Studies by psychologist Robert Cialdini have revealed that social norms are powerful means of encouraging people to do all sorts of things. In an experiment on how to cut down on washing hotel towels, for example, when guests were told that most other guests staying in the hotel reused their towels at least once during their stay, and therefore saved water and energy, they were 26 per cent more likely to do the same than if they were told only of the impact on the environment of daily towel washing.[2] Cialdini noted that if guests were told that the majority of guests 'who stayed in this room' reused their towels, the impact was even greater. The most powerful motivator is the suggestion that *other guests just like them* were doing it.

Making us behaviourally commit to something

Towel-washing is a nagging issue for many hotel chains because of the toll it takes on water and power. The Disney Hotel in Anaheim, California, wanted to encourage towel reuse and tried a commitment bias-focused experiment among its guests. Some guests were asked to pledge to reuse their towels during their stay and were rewarded with a Friends of the Earth badge if they agreed to do so. Other guests (not specifically asked to reuse) acted as the control group for the experiment. After a number of weeks

– during which time 2,000 guests had stayed at the hotel – Disney analysed the results. They found that guests were 25 per cent more likely to reuse their towels when they had pledged to do so and had received a Friends of the Earth badge, and they also hung up over 40 per cent more used towels compared to the control group. Researchers estimated that the impact of the pledge and badge combined would save the hotel 2,500 loads of laundry, and therefore 700,000 gallons of water and $51,000 per year. The researchers noted that pledging to reuse towels had a bonus advantage, with an approximately 70 per cent chance that pledging guests would also turn off the lights in their rooms, compared with only a 50 per cent chance among guests in the control group.[3]

Making choices for us

We often accept what is put in front of us and are happy to go with the flow, but this means accepting a choice that has been made for us – the default option. Companies often structure the choice architecture – the context in which our options exist – to ensure that we select that default: the option they want us to choose.

As part of a Corporate Social Responsibility initiative, Disney altered the children's menus at its theme parks from 2006 so that healthy items became the default option. Remember the status quo bias? We don't like change, and sometimes we are too lazy to change things, so we stick with the default setting in lots of contexts. Here Disney flipped the default to aim healthy and to save us the responsibility of making the responsible choice.

Whereas previously the default kids' drink had been a fizzy one and the default vegetable side dish had been French fries, the *revised* default drinks became healthier choices such as pure fruit juice, water and low-fat milk, and the default side dishes with children's meals became vegetables rather than fries. When guests order a kids' meal including a drink and side dish, it now automatically includes healthier options. These changes have been successful; the majority of families stick with the healthy children's meal defaults rather than requesting the less healthy choices. Only in Paris, where there was a markedly unenthusiastic (15 per cent) uptake rate for healthy side dishes, did the French stick with French fries![4]

Software guardians McAfee run a tight ship when it comes to keeping an eye on your computer security, as I found out recently. Woe betide you should you let your subscription expire – not on their watch. After a series of on-screen panels had repeatedly popped up to tell me I was no longer being protected by them (because I had signed up with another supplier), including click-options to 'RENEW' or simply close the panel, they brought out the big guns, delivering a rather threatening choice architecture. I could either '*Reactivate Protection*' or I could '*Accept Risk*'. The default choice is highlighted in a white click-box, while *Accept Risk* is contained within an alarming-looking red one. If I refuse to go with the choice they are suggesting, I have to click *Accept Risk* and accept also the anxiety and negative priming of the 'risk'. (The red pop-up panel also depicts a shady figure hunched over a laptop right next to the *Accept Risk* button. Is this McAfee's plan? Do they really want to scare the bejesus out of me?)

Using anchors or reference points to direct our choice

Anchoring is a particularly popular tool often used by companies to set up a choice architecture that guides you towards making a particular choice. It's a good name for this bias because it means we make a mental attachment to anchor to one particular point and then assess our options and adjust from that point. Add to that the human instinct to head for the safety of the middle ground – not too hot, not too cold, but somewhere in between will be *just right*.

Go to Oxfam's website and click on the 'Donate' page, and you will be offered three choices of monthly donation levels: £5, £15 or £20.[5] Assuming you're in the frame of mind to sign up, it's more than likely that you won't opt for £5, which might seem a little ungenerous, while £20 a month is too much of a commitment. You anchor on the £5 (too little), glance at the £20 (too much) and can't help but deem somewhere around £15 just right. It's a good example of another bias: extremeness aversion, which makes us uncomfortable with extremes, preferring the 'just right' middle ground (though perhaps it should be renamed the Goldilocks Illusion). Oxfam fully engages our likelihood to anchor on the middle option too; the option to pay £15 is the one they highlight.

If you've ever been asked to donate on the JustGiving site, or something similar, you'll have found yourself anchoring there too. How much to donate? What did the last person give? What did your friends give? What did the parents of the fundraiser give? These are significant anchors. We're always looking for something to hang on to.

Recently, London museums and galleries with free entrance policies and an optional donation system have powered up the 'donation' element by incorporating a series of head-on anchor nudges. Visitors to the Science Museum and the British Museum can get into the museums only after passing through a manned 'welcome' desk where staff suggest they make a donation. In the Natural History Museum, visitors enter via 'Donation' card-tap points, where at each point you can tap for either £5, £10 or £20. With the £20 point being the top anchor (the decoy?), the aim is to encourage the majority to give £10. (The Goldilocks Illusion again – £5 too little, £20 too much, £10 just right.)

Booking.com too makes us feel as if we are getting a good deal by quoting the standard price, anchoring us to that price, and then offering us a cheaper quote, cementing the perception of the excellence of the deal by adding in many seemingly 'free' items such as breakfast, Wi-Fi and penalty-free cancellation. It also tries to counteract our tendency for procrastination and inertia by impressing on us the speedy booking process with tags such as *'Book now! It only takes 2 minutes!'* From a personal perspective, it really works – you can get yourself into quite a frenzy making that booking before the chance goes (don't forget, there are at least eleven other people looking at this hotel *right now* and only one room left…).

We've seen how, in a restaurant, the choice of a bottle of wine is often based on anchoring somewhere on the cheapest and, progressively, the more expensive bottles, and then selecting something usually one or two above the cheapest. Think about how you judge shops, restaurants, holiday companies and so on, and the chances are you will search for some form of anchor that will tip you towards one rather than another.

Here's a great example of how adding a canny dash of behavioural science to customer service can transform a bad experience (in this instance, the problem of a cancelled flight) into a good one, by a clever use of anchoring. Say you need to be in New York tomorrow morning but your 11 a.m. flight is cancelled. There's an evening flight that's open, though. Where some airline reps might simply say, 'I can put you on the flight leaving at 9 p.m. tonight,' others, knowing full well the 9 p.m. flight is available but seeking to achieve the maximum goodwill from the customer's reaction, might say, 'Well, I know I can definitely put you on the 7 a.m. flight *tomorrow*, but let me see what I can do to put you on the earlier flight, which is at 9 p.m. *tonight*, as I know you really need to be there in the morning.' Offering the prospect of a much less desirable option *first* creates a mental anchor, making the best alternative seem far more satisfactory than it would otherwise have been if it had been the *only* option. Rather than being irritated that the 11 a.m. flight was cancelled, you're probably delighted that the rep has apparently gone out of their way to 'secure' a seat for you on the evening flight. A difficult situation with a disgruntled customer has become a successful encounter with a satisfied customer. A very clever switch.

On a daily basis, as customers, we'll find ourselves influenced by anchors, whether it's the cheaper 'SALE' price in red next to the crossed-out original price of an item, or the promise of 'up to 70% off', or high-to-low suggested amounts to donate on a charity website. Our brains are recalibrating and recalculating, without conscious deliberation, a lot of the time. Sometimes we have to take a breath and realise that, while that sale price can be incredibly motivating – down from £6,740.00 to £999.99, for instance – we don't really need that Versace wedding dress on the TK Maxx online store.

Using visuals or symbols to subconsciously prime our behaviour

Visual priming packs a powerful nudge. Here's a quick demonstration. Put the same amount of food on both a large plate and a smaller plate, and the amount of food on the smaller one will appear more generous and be more likely to fill you up. It's the standard trompe l'oeil to use if you're on a diet.

When a cafeteria presented food on smaller plates in this way, people ate 22 per cent fewer calories. And when the same restaurant arranged fruit in an attractive bowl rather than jumbling it together on a catering tray, fruit consumption increased by 103 per cent.

Enlightened nursing homes and hospitals use what might be termed compassionate visual priming. Staff in a hospital in Leicester place the image of a butterfly on the door when a patient requires peace and quiet or may be in the last stages of life. The image of a butterfly signifies fragility, beauty and transformation, and primes those entering the room to act gently and quietly.

Research has shown that symbols are processed fast and intuitively by our System One thinking, a particular bonus for time-starved, hard-working nursing staff who don't want to exert too much of their precious energy interpreting information.

The Hospice Friendly Hospitals programme is the initiative of the Irish Hospice Foundation. It developed the End-of-Life Symbol, which is displayed when a patient has died and after the family has been informed.[6] Awareness of this profound event allows staff to interact appropriately with those affected by the death.

In some care homes where elderly patients are suffering from dementia, photographs on the doors to their rooms showing them in their youth, perhaps on their wedding day, looking happy, active and young, prime staff and visitors to think of them as they were in their heyday.

It was the power of symbols to communicate fast that led to the development of the bicycle lights Brainy Bike Lights. The illuminated area was in the shape of a cyclist, the aim being that the symbol communicated 'cyclist' very quickly to other road users and primed them to intuitively respect the fragility of a human being on the road.

Brainy Bike Lights explained how symbols work hand in hand with System One thinking:

> Well-known symbols are processed fast and intuitively by what scientists call our System One thinking, so we already knew that a bike light based on a bike symbol would create a simple and clear warning message for drivers versus a traditional red or white bike light. Our bike symbol is effectively a cognitive shortcut.[7]

Here are two more, very different examples of visual priming, each being used to address anti-social behaviour. The first, a striking idea from the UK, was introduced to discourage vandals from damaging a row of shops in Woolwich in south-east London that had been attacked during the summer riots of 2011. A local group of graffiti artists painted images of babies and young children on the security shutters that covered the shopfronts; the images were based on photographs provided by local families. The simple priming idea was that, when the shutters were down, they revealed a gallery of portraits of neighbourhood youngsters whose chubby cheeks and big eyes triggered protective and caring instincts to come to the fore and thereby discouraged vandalism.*

The second is perhaps one of my favourite examples of organisational nudging and comes courtesy of the men's urinals at Schiphol Airport, Amsterdam. Urinal users, it turns out, are not particular about aim, which makes cleaning the floors of the men's toilets time-consuming and costly. The airport authorities wanted to cut costs. Should they add signs asking men to refrain from peeing on the floor? The airport manager Aad Kieboom borrowed an idea from Jos van Bedaf, manager of the airport's cleaning department: rather than putting up signs warning against misdirection onto the floor, he simply had a small fly engraved on the urinals to encourage users to take direct aim. It's thought that the fly has reduced floor spillage by about 50 per cent and cleaning costs by 20 per cent. A simple but effective solution, and a great example of 'small change, big difference'. A new graphic has recently been added to Schiphol urinals, harnessing the same need to aim for a target: a flagstick on a golf course.

With your eyes wide open, you should now be much more aware of how most commercial companies are activating the brain wirings from Section 1 to influence our thinking and behaviour 24/7.

So, how are governments using brain insights to nudge and steer us? (Upfront disclaimer: you might find the stuff in this section a tad on the dry side, but it's no less relevant for all that. Governments are using behavioural science to get you to behave the way they believe is the best way.)

---

* The project, called Babies of the Borough, was initiated and paid for by advertising agency Ogilvy & Mather.

Over the last few years we have seen governments around the world at the forefront of the application of these insights to coax citizens in the direction they'd like them to go. From a government perspective, one of the attractions of the application of behavioural science-led techniques is that they are all about changing citizens' behaviour and choices, as opposed to a more complex and often more time-consuming attitudinal change, which may not lead to real behavioural change anyway.

This part of section 2 will explore the application of these brain insights in four key areas:

⊙ Health
⊙ Politics
⊙ The environment
⊙ Good citizenship

In the UK, the Cabinet Office works with a group of advisers called the Behavioural Insights Team (BIT). The BIT was established in 2010 as a 'unit embedded in the heart of the UK government',[8] which has now evolved into a 'social purpose company' partly owned by the Cabinet Office, NESTA (National Endowment for Science, Technology and the Arts) and the BIT's own staff. The BIT's remit is to apply insights derived from the behavioural sciences to policy and communications.

Things have moved on apace since 2010, though. Many other areas of government have also invested in behavioural insights. There are teams in the Department of Health and Public Health England, and in other departments, such as the Department for Business, Innovation and Skills, the Department for Work and Pensions and the Department of Energy & Climate Change. According to the 2016–17 Update Report from the BIT, 'In the UK, almost every major government department now has a behavioural insights function of its own.'[9]

But understanding and applying the behavioural sciences is not the preserve of the UK. There are government 'behavioural teams' across North America, Latin America, Europe, Asia and Australia. International organisations of the calibre of the Organisation for Economic Co-operation and Development (OECD), the World Bank, the United Nations Development Programme and the European Commission have all embraced the behavioural sciences. In fact, according to the OECD, there are 196 institutions around the world applying behavioural insights to public policy.

In 2017 the OECD published a report detailing over one hundred case studies where behavioural insights had been applied in this way.[10] They included:

⊙ reducing water consumption in Costa Rica
⊙ keeping young people safe in South Africa
⊙ fighting antibiotic resistance in the UK
⊙ encouraging the use of ashtrays on the streets of Denmark
⊙ screening for diabetes during Ramadan in Qatar
⊙ tackling obesity by rewarding exercise with air miles in Canada.

Even back in 2011, just a year after it was founded, the BIT delivered insights into a wide range of challenges, including:

⊙ how to increase truthfulness among taxpayers by using the commitment bias more effectively
⊙ how to harness the power of social norms/herd instinct to save money
⊙ how changing a default from passive to active choice can have a radical impact
⊙ how using a combination of bias-driven approaches can build impact and efficacy
⊙ how applying social norms (the herd instinct again) can speed up response times for surveys

And even, potentially:

⊙ how to reframe a tax payment as a potential windfall!

Nudging and steering our behaviour in the healthcare context

The UK government has used behavioural science to have a direct positive impact within the National Health Service in terms of patient communications and – increasingly key these days – in relation to saving money. Here's how behavioural biases have been applied to save money and to nudge patients and doctors to do the right thing.

Discouraging patients from missing appointments

One clever use of behavioural science has been applied by a number of hospital outpatient departments and general practices in the UK. It focuses on missed appointments: between 2007 and 2008 there were 6.5 million 'did not attends' (DNAs) for hospital outpatient appointments in Britain, with hospitals losing around £100 every time a patient did not show up. (That's £650 million lost!) GP surgeries suffer in the same way – each year patients miss around 10 million GP appointments and 5 million appointments with a practice nurse. A couple of practices in the South of England have experimented with behavioural steers to tackle the problem. In combination, the two practices were making 10,000 appointments each month, of which the number of DNAs was a staggering 4,700. They adopted three behavioural steers:

- They asked patients who made an appointment over the phone to repeat back the appointment information (date and time) to the receptionist, priming them to turn up – a form of commitment bias.
- They asked patients who made an appointment in the surgery to write down the appointment information on a card. Writing the information for ourselves engages us in the agreement and means we'll feel compelled to show up – commitment bias again.
- They displayed a large sign in the waiting room stating that the vast majority of patients turned up to their appointments on

time, with the implied social norm nudge that since most other people do this, so should you.

Working together, these approaches succeeded in improving patient attendance by a third – and counting. If hospital outpatient appointments were managed in the same way, there would be more money for healthcare. In fact, it seems hospitals may have already started to follow suit. Accompanying a recent letter to a colleague about a physio appointment was a leaflet highlighting the cost to the NHS in millions of pounds when patients failed to attend. You may have experienced these nudges first hand yourself.

In hospital waiting rooms, posters give a more personal toll to the lost millions of pounds caused by those missed appointments, translating them into numbers of much-needed nurses and doctors, to additional beds and equipment. Powerful messages all.

Encouraging doctors to prescribe generic drugs, i.e. saving lots of money

Money can be saved in drug-prescribing too. GPs can choose between generic (cheaper) and branded (more expensive) drugs, and do so using an online system that will be very likely to present the more expensive drugs at the top of the list. Time pressures and an assumption that if a drug is mentioned first it's because it's the most frequently prescribed – and therefore must be good (think social norms) – can result in GPs prescribing the pricier (default) option. Fortunately, in the NHS generic prescribing has been on the rise since the 1970s and stood at 84 per cent in 2015. The King's Fund estimates this has saved the NHS around £7.1 billion in total. However, it calculates there is still room for improvement, with potential for rates to rise to 90 per cent, especially as there is variation between general practices.[11] In the US, generic drug prescription rates stand at 89 per cent, but again they could be higher. In a society where the payment often falls on the patient, generic use plays an even more important role since patients are nearly three times more likely to abandon medication altogether if is branded, because of the high cost.[12]

In the United States, Mitesh Patel, Director of the University of Pennsylvania's Penn Medicine Nudge Unit, recently trialled a tiny tweak to the prescription order system on the university's electronic health record system.* When doctors came to select the drug they wanted to prescribe, they clicked on a drop-down menu. Previously, branded drugs were listed at the top of that menu and generics at the bottom. Patel flipped the order so that generic drugs were listed first. It had an astounding effect. Before the trial, the generics prescribing rate at Penn Medicine was around 75 per cent. Immediately after the change in the drop-down order, the generic prescribing rate leaped to 98.4 per cent, then remained there for the duration of the ten-month test period.[13]

Why might this tiny change have had such a big impact? Sometimes the order in which items are listed can have subconscious effects on our decision-making. For instance, it may be that doctors presumed the medical community's preferred choice of drug was the one listed first. Or perhaps, short of time and energy, they scrolled down to the first drug they saw that matched what they were looking for and looked no further.

As Patel concluded, this change required only a minor modification to the prescription order system and took very little effort, but it will probably save millions of dollars over the next few years for patients, insurers and the health system. While Patel admits that one of the biggest barriers to these kinds of interventions is that many clinicians are resistant to change and to the concept of being 'nudged', he counters that 'we often

---

* The Penn Medicine Nudge Unit was launched in 2016 to systematically develop and test approaches drawing on behavioural science to improve healthcare delivery. It is the world's first research unit of its kind within a health system.

don't realise that we are already being nudged by the design and choice architecture of whatever electronic health record system we are using. It influences our choices every day, but often this is overlooked.'[14]

Preventing the highly damaging overprescribing of antibiotics by doctors

Behavioural nudging has also delivered some good results in the battle against the overprescribing of antibiotics, which is becoming a huge problem, with the increasing number of strains of antibiotic-resistant bacteria. In the United States, 41 million antibiotics prescriptions are written each year for adults, yet half of these are unnecessary and inappropriate.

A group of behavioural scientists set out to see if they could use commitment bias to reduce unnecessary prescriptions. They took a group of fourteen doctors working in community health centres in Los Angeles and asked them to sign a letter stating their commitment to avoid any injudicious use of antibiotics. Their signed letter was then displayed in their examination room for all patients to see, making their commitment highly public. People like to be consistent with their behaviour and do what they say they will do. What's more, publicly committing to that behaviour links it to self-image and can actually increase dedication to carrying out that behaviour.

During the nine months before the initiative, around 43 per cent of prescriptions were unnecessary. The initiative then ran for twelve weeks. While a control group (with no commitment bias intervention) saw an *increase* in inappropriate prescriptions to nearly 53 per cent, the commitment group saw a reduction to 33 per cent. Apply this level of reduction across the entire United States and the commitment letter could stop 2.6 million unnecessary prescriptions, save $70 million annually and tackle the problem of antibiotic-resistant strains of bacteria. Another incredibly simple change – with a dramatic impact.

Encouraging patients to take their medication

In the US, patients' failure to take their medication as directed contributes to 125,000 deaths and wastes around $290 billion *each year.* For a variety of reasons (often to do with side effects or being 'symptom free'), people with serious or chronic conditions (heart disease, high blood pressure, high cholesterol or diabetes) often let their prescriptions lapse or simply fail to take their drugs.

In the UK, 2.9 million people are diagnosed diabetic, and it is estimated that 35 per cent of the population is pre-diabetic. Taking medication, following a strict diet and executing lifestyle changes can all improve the health of patients with type 2 diabetes. However, patient 'adherence' (defined as the extent to which a person's behaviour corresponds with agreed recommendations from a healthcare provider) is estimated at 50 per cent or below. A recent behavioural intervention study, run by The Behavioural Architects with Hall & Partners, delivered impressive results with seven out of ten patients showing increased adherence and 80 per cent with total compliance over the intervention period. In the study, patients were given a behavioural science-inspired action poster to put on the wall at home; they also made and signed a contract promising to take their medication and the contract was signed and witnessed by their supporters (textbook commitment bias). The poster included images of a member of the patient's family or someone else close to them to increase the significance of the communication, and they were given daily stickers to keep a tally of their adherence and record their journey step by step (chunking).

Encouraging people to sign up to organ donor schemes

In May 2020, the DVLA introduced a change to the organ donation scheme whereby, unless new driving licence applicants actively opt *out* of organ donation, they will be presumed to have opted in. The wording on their website reads as follows:

> It will be considered that you agree to become an organ donor when you die, if:
> ⊙ you are over 18;
> ⊙ you have not opted out;
> ⊙ you are not in an excluded group.

While 80 per cent of adults express willingness to donate their organs, only 40 per cent are currently registered donors. Similar opt-*out* schemes operating in many European countries have donor registration levels of around 99 per cent, demonstrating how the default setting, in this context anyway, can actually be a good thing.

Encouraging councils to nudge healthier lifestyles at a local level

In 2016 the Local Government Association published a document called 'Behavioural Insights and Health' that aimed to show how behavioural insights can be harnessed to encourage people to live healthier lives. What the document has achieved through nudging and steering is rather wonderful. London's Hounslow Council embraced the science wholeheartedly, applying it to healthier and more environmentally friendly choices by encouraging the following:

⊙ children walking to school: they sent maps and travel advice about popular and safe routes to parents of children about to start school and instigated a 'Beat the Street' inter-school competition that allowed children to earn points for their schools by tapping personal cards into card readers as they arrived at school

⊙ dental visits: they sent first birthday cards to children that included a message encouraging parents to take their toddlers to the dentist

- ⊙ healthier eating: they focused on traffic-lighting foods to show the healthiest and moved these closer to tills in shops in civic centres
- ⊙ best recycling practices: to coincide with a house move

Nudging and steering our behaviour in a political context

As voters – be it in general or local elections, or referenda – we are rarely left to our own devices to make up our own minds. A lot of money and campaigning goes into trying to make an impact on our decision-making.

Encouraging voter registration and voting

One of the most exciting first large-scale uses of nudging through behavioural biases took place during the US presidential elections in 2008. Barack Obama's campaign team knew that the higher the turnout at the polls, the more chance Obama had of winning the election. About two weeks before Election Day, Obama's team got together an advisory group of twenty-nine of the nation's leading behaviourists, including psychologists and economists, to devise a sure-fire way of getting people to the polls.* The campaign team fed all communication channels with one message: 'A record turnout is expected.' The behavioural science-savvy team knew that once a commitment to vote had been leveraged, the most powerful motivator for action would be the social norm suggestion that everyone was doing it. 'People want to do what others will do,' says psychologist Robert Cialdini, author of *Influence: The Psychology of Persuasion*.[15] It's a piece of pure behavioural genius: most people don't want to be left out and they'd prefer to be on the winning team. And it worked, inducing some 5 million people who usually

---

*   The team is believed to have included many of the people whose work is referred to in this book: Dan Ariely, Richard Thaler, Cass Sunstein and Daniel Kahneman.

did not usually vote to head to the polling stations. When the votes were counted, Obama had 52.9 per cent of the popular vote and 365 electoral votes, overwhelming the 270 total he needed to win. After being elected, Obama established a government-based behavioural insights team and was champion of its power throughout his two terms in office.

In the 2012 presidential election, the team created a multi-layered behavioural approach, building on the successful 2008 campaign and adding other behavioural science concepts, such as commitment bias, to ensure people would vote. Time was productively spent on doorsteps with a three-pronged commitment bias message in which:

- canvassers made it clear they were familiar with the voter's past voting activity, saying: 'Mr X, we know you have voted in the past...' on the basis that most of us like to stay true to our past behaviour to avoid a conflict of identity.
- they asked people if they would sign a 'Commit to Vote' card that carried a photograph of Barack Obama, for them to keep and display.
- they asked whether people had made a specific logistical plan to vote, and if they had not yet done so, to make one, including the time of day they planned to visit their polling station.

Once again, Obama triumphed, winning 51 per cent of the popular vote and 332 electoral votes.

Nudging and steering our behaviour in encouraging environmentally favourable behaviour

It is in our best interests to behave in an environmentally friendly way and much effort is made to manage our behaviour accordingly.

Nudging people to cut down use of plastic bags by changing the default

Not so long ago we all went empty-handed to do our supermarket shopping, loading our stuff into branded plastic carrier bags at the checkout and then, at home, stuffing the emptied bags into broom cupboards and drawers to keep, until finally throwing them all away in a massive clear-out. The realisation dawned that all these plastic bags were non-biodegradable and caused environmental damage – yet in 2014 British shoppers took *8.5 billion* of the things home from major supermarkets.[16] The UK government took it upon themselves to change the default. In October 2015, they imposed a 5p charge on each single-use plastic bag used and supermarkets removed the bags from the checkouts. The result? The use of plastic bags by shoppers was reduced dramatically. Government figures show that the total number of single-use plastic bags almost halved in two years, from 2.12 billion in 2016–2017 to 1.11 billion in 2018–2019.[17] It is now normal behaviour to take your own bags to the supermarket; it has become the default setting. Some supermarkets even reward you for using your own bags – a little bit of reciprocity bias.

Nudging and steering our behaviour in encouraging what is perceived as good citizenship

Perceived good citizenship is subjective, but can involve anything from filling in a census to paying taxes on time to voting in elections.

Priming people to be more honest by harnessing the commitment bias

We are all familiar with filling in official forms. We know that at the end of the form there is always a 'truth' declaration to sign and, let's be honest, sometimes at the point of signing you can feel a little uncertain: *Have I definitely declared everything to the best of my knowledge? Is there anything at all I've accidentally omitted to divulge?* But you rarely revisit the entire form just to make sure; you sort of hope for the best.

An experiment in the US found that asking people to sign a declaration of truth at the *beginning* of their car insurance application led them to go on to declare 10 per cent higher mileage.[18] This may be because once you've said you're going to tell the truth (with the suggestion perhaps of 'the whole truth and nothing but the truth') people feel more committed to continuing to tell the truth when answering subsequent questions. The upfront declaration effectively primes us to be truthful. If we have to sign a form at the end, when we may have been only 'partially accurate' in answering questions, correcting the form means admitting that we have been dishonest to start with. The idea of priming honesty in this way is still under examination and there have been conflicting research results and some lack of replication in studies. It's a case of 'watch this space'.

In *Nudge, Nudge, Think, Think: Experimenting With Ways to Change Citizen Behaviour*, Peter John, a leading professor of politics and economics, and an adviser to the BIT, said:

> It's about working out which buttons to press. People generally like to conform. People also seem to associate signing a form at the top with signing a legal declaration, which is more serious than signing it after you have filled everything out.[19]

Leveraging social norms to get us to pay our taxes more promptly and complete the Census form

A trial organised by the BIT and Her Majesty's Revenue and Customs (HMRC) in early 2011 to encourage tax debtors to pay up used a social norm statement to hit the message home. Here's an assessment of the results from the Behavioural Insights Team's annual update that year:

> The trial was on a large scale, comprising around 140,000 debts worth £290 million. The results were that letters which informed people that the majority of people in their area had already paid their tax, and which reminded people about the importance of paying tax for their local services, outperformed the control group letters by around 15 percentage points. That is a significant result, which we will be looking to apply in other areas of debt, fraud and error. We estimate that if the most successful letters were sent to all self-assessment customers, and the tax collector resource freed up were used to bring in other uncollected Exchequer debts, it would generate £30 million of extra revenue to the Exchequer annually – as well as advancing over £160 million of cash flow by approximately six weeks each year. HMRC's wider self-assessment debt campaign increased cash-to-bank by more than £350 million in the first six weeks of the campaign, nearly treble the amount collected during the same period last year.[20]

Whether we like to admit it or not, we don't like to stand out from the crowd. In 2011 the UK government took advantage of people's fear of not being part of the herd, by mailing Census letters that included phrases like 'As you are one of the last in your area…' to chivvy people into completing their forms. The letters implied both the failure of the Census slouches to be good citizens (completing the Census is a civic duty, after all) and highlighted the fact that they were in the minority.

Encouraging us to spend more too

In the US, following the 2008 financial crisis, the Obama administration was looking to boost consumer spending. It designed two tax rebates that totalled more than $235 billion in 2009. Yet with behavioural science in mind, the rebates were delivered in small amounts through monthly pay cheques rather than in single large lump sums. The theory behind this rests on a bias called mental accounting. We haven't mentioned this one yet, so here's a quick explanation. Mental accounting is the term given to the way people treat money differently depending on how they came by it and on what its intended use might be. For instance, money that it is earned is likely to be treated differently from a windfall as the result of a lottery win. If you happened upon a £20 note in the street, you would be more likely to spend it frivolously than if you had got the money by working for it. The size of the rebate matters: the bigger the rebate the more likely it is to be ring-fenced and saved for a major, but possibly future, purchase. It was recognised that if the amounts of money were small enough, people would not notice them so more would be spent and the economy would benefit.

The success of this rebate strategy and the behavioural science behind it was confirmed in a *New York Times*/CBS News poll in September 2010 which revealed that fewer than one in ten respondents knew that the Obama administration had lowered taxes for most Americans.[21] Half of those polled said they thought that their taxes had stayed the same; a third thought that their taxes had gone up; and about a tenth said they did not know.

How China has taken the application of social norms to the extreme

China seems to have opted for more of a Big Brother-style approach than other governments. Its 'social credit system', first launched in 2014, is based on the premise that 'keeping trust is glorious and breaking trust is disgraceful'. The system awards citizens a base score of 1,000 points and then adds and subtracts points for each good or bad deed committed. You can lose points for driving misdemeanours, failing to pay bills or taxes on time, playing too many video games or posting fake news; you can win points by making charitable donations, acting heroically, helping your family in tough circumstances or having a successful business.

In Rongcheng, an early pilot city for the scheme, boards were displayed explaining how you could gain or lose points and showing social norm-driven photographs of high scorers. While winning points can ensure ease of travel, discounts on heating bills and favourable bank loans, those who fall foul of the system can find themselves on a blacklist that prevents them from buying plane or high-speed rail tickets. Dr Samantha Hoffman, a senior analyst at the Australian Strategic Policy Institute, describes it as Orwellian: 'It's a pre-emptive way of shaping the way people think and shaping the way people act,' she says.[22]

These powerful applications of behavioural insights around the world, and the very existence of the Behavioural Insights Team in the UK, demonstrate the culture of experimentation and the openness to understanding and applying new learning on behavioural insights. This also suggests there will be more applications in future government communications and campaigns. Keep your eyes peeled.

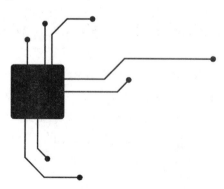

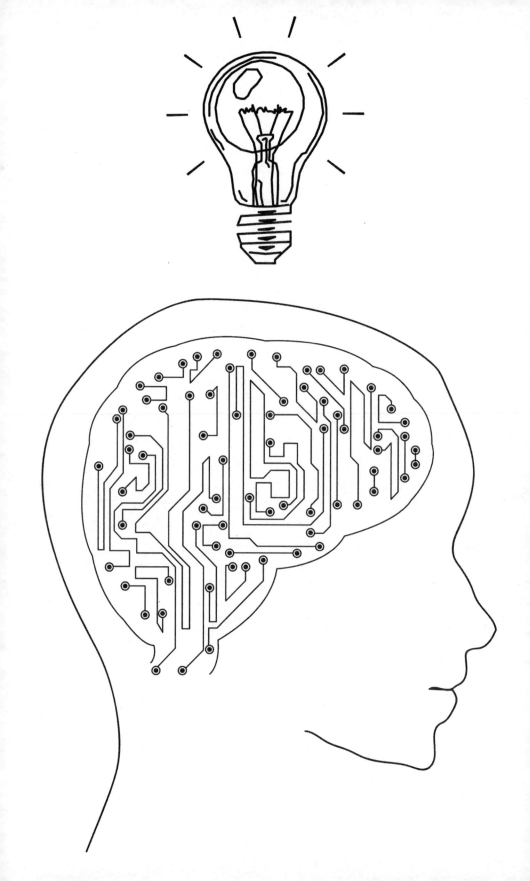

Section 3: Taking action: Simple ways to rewire your brain

Getting practical

We've examined the power of a number of different behavioural biases and shown that they are all around you, nudging and steering you towards one decision or another. Your awareness of them should be heightened now, so it's time to look at how you can take charge of them – either by consciously applying a bias to a specific aspect of your life, or by confronting or rejecting its influence.

To inspire you, we are going to learn from the experiences of members of the Brain Team, described in the Introduction. Through their personal journeys, you will see how an increased consciousness of the wirings enriched their lives in simple ways and how this increased consciousness can change your life too. This section is packed with ideas that you can put into practice and it will show you that even the smallest of changes can have a big behavioural impact.

In what was effectively a live behaviour-change experiment, our wonderful Brain Team participants embraced five biases and two brain strategies: autopilot, reciprocity, status quo, egocentric, normalising, and the brain strategies framing and chunking. They focused on them, one by one, for a week at a time, and recorded how their lives changed when they responded to the influences of each bias with conscious understanding and awareness.

Using the direct day-to-day experiences of the Brain 'Team', we will examine the impact of the biases and strategies on the participants and see what happened when they understood what was going on and exerted a more conscious 'control' over the brain wirings. We will:

- ⊙ explore our dominating autopilot and find out what can happen if we switch it off, even if only for a moment
- ⊙ discover how easy it is to start positive behavioural waves and stop a negative backlash by playing to the brain's in-built reciprocity
- ⊙ increase our awareness of how we are wired to like things the way they are, to keep the status quo, and what happens if we rock the boat occasionally
- ⊙ acknowledge and confront our egocentric life filter
- ⊙ challenge normalising processes and find out how to stop taking things – and people – for granted
- ⊙ experiment with changing mental frames and see how this can affect the way we interpret or reinterpret information so we can set about achieving tasks and solving problems more positively and capably – change the frame, change the meaning
- ⊙ tackle breaking issues down into manageable mental chunks to make our goals easier to achieve.

We will show how you can use this new awareness not only for short-term, temporary changes but also for longer-term, permanent changes in behaviour.

Here's the structure. For each bias and strategy we will examine:

- Brain-wiring insight – how it works on you
- How it operates in our lives – why it has the influence it does
- Your brain-rewiring plan – how to combat/manage its influence
- Brain-rewiring reward – the benefits of taking charge
- Specific nudges for dealing with the bias or strategy – practical ideas you can actually use
- Reflections from the Brain Team (in italicised quotations)
- Brain-rewiring awareness
- Summary of brain nudges

So, let's start with a little behavioural priming for you:

DON'T be put off if the brain insights sound like common sense or if you feel you are already familiar with them – familiarity is a very different thing from application or confrontation. The most powerful insights are often the ones that seem obvious or echo things we already know. It is the heightened consciousness of the insights, followed by their application to daily life, that triggers the magic. Even the smallest responses to these insights can give you pause.

It's been a really positive experience and the exercises have been practical, and with immediate pay-offs. Go into this open-mindedly, and suspend your cynicism and give it an honest try. You'll reap the reward of happier days!

DO be open to the process and you will see how deepening your consciousness of these various brain wirings will take you on an enriching journey during which you may well learn from, and be inspired by, the experiences, actions and reflections of others.

The insights made me smile, find goals achievable, love my children, love my husband, love the countryside, love a quiet time, so many things.

DON'T worry if some of the insights engender an element of fear, because it's only natural that they might; our lives are our lives after all, and we have enough to do to live them without being asked to change; change is not always a welcome prospect. Just relax: nothing is set in stone; there is no *one* way of doing something. The first stage is simply awareness, and just thinking about the insight, or how your brain is wired, can prompt changes. As in yoga, simply thinking about a movement can effect change; it's a conversation you can have with yourself, a question you can ask and ponder. Then, and only if you want to, you might decide to try something or do something differently – but it's you who is in control. It's you who decides if you want to do something, and what and when that might be. And remember, a small change really can make a BIG difference.

Over the next six parts you will see how other people found ways to gain new self-awareness and new mindfulness. You'll discover:

⊙ how having your eyes wide open may sometimes reveal home truths that are difficult to accept
⊙ how all of us can benefit from a little self-observation every now and again (it's an opportunity to consider and reflect on how we navigate our daily lives)
⊙ how we can learn to refocus on what is important
⊙ how we can affirm our strengths as well as acknowledge our weaknesses
⊙ how we can improve the quality and strength of our relationships
⊙ how we can deepen and enrich conversations, and give ourselves the gift of viewing things from new perspectives

Overall, you will learn how to make a big difference to your life, and to the lives of those around you, by making only small adjustments.

Let's begin the journey with some advice from a Brain Team member:

The chief advice I would give to others is to try it, stick with it, participate fully and feel the insights and changes. The reason being that, if you do, the chances are very high indeed that in looking back you will realise that by granting yourself permission to focus on yourself (often difficult for people with constant demands upon their time) you will see just how far you have come and what huge gains you have achieved. And the beauty of it is that it need not stop there but remain a template for continuing to improve and bring about profound changes in not just your own life but other lives too. In short, what have you got to lose?

AWARENESS

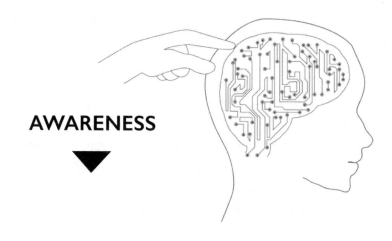

UNDERSTANDING

ACTION

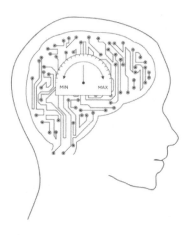

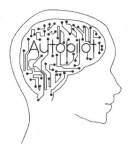

1. Autopilot

We're wired to be on automatic: How to shift to manual control

Autopilot brain-wiring insight

Most of what we do is on autopilot. We gave our Brain Team the following autopilot insight to ponder over the course of a week, inviting them to maybe turn their autopilot off occasionally, and see what happened:

> Approximately 95 per cent of our behaviour is carried out without conscious thinking. This is, of course, brilliant because there is no way we could (or would) want to think about everything we do all the time, or to process all the information presented to us at every moment. But there is a downside to being on autopilot, because it can also mean that life slips through our fingers a little quicker than we might like. So, what I would like you to try to do is to turn the autopilot off. Try to break from a few of those things which have become a daily routine with little active conscious thought. Try to have your eyes wide open rather than eyes wide shut.

How autopilot operates in our lives

The iceberg metaphor (discussed in Section 1) explains how our conscious thought makes up only a small portion of our minds; most of our thinking (like the part of the iceberg that's under water) is carried out unconsciously. So, whether we know it or not and whether we like it or not, we are operating for the most part on autopilot. In lots of ways this is a good thing – we don't need to think about everything we do all the time, or to process all the information presented to us at every moment (and there is no way we would or even could). We don't think about loads of things we do on autopilot, because we don't have to: we breathe without having to galvanise our lungs into action and speak our mother tongue instinctively, without having to preform sentences in our heads and try them out for accuracy; we chuck a couple of pieces of bread in the toaster without running through the process of reasoning why this might be a good idea; we brake, change gear and steer our cars without thinking about moving our hands and feet; we don't have to work on any conscious level to do these things and that is what our autopilot is there for.

But there is a downside to being on autopilot, because it can also mean life slips through our fingers a little quicker than we might like. Things become habits and it's easy not to think about whether it might be good to do them differently. Autopilot helps you to survive but it does not help you to experience life fully.

A study conducted among 5,000 adults by the Child Accident Prevention Trust in 2011 made these disconcerting, but not terribly surprising, discoveries:

⊙ Half of the participants said they travelled to work on autopilot.

⊙ A quarter of participants were unable to recount the exact details of their commute on the day they were interviewed, including whether they had stopped at red traffic lights.

⊙ Half of them had turned back unnecessarily on a journey because they could not remember if they had locked the front door.

We depend on our autopilot ability, but we also need to be aware just how much we use it. There is evidence to suggest that the more habituated and routine our lives, the less chance we have of creating new memories. It's something to think about, because the older we get, it's likely that we'll have fewer new experiences – unless we go out of our way to avoid routine and monotony. The concept of repetitive patterns diminishing our lives is expressed perfectly in this extract from a book by science writer and journalist Joshua Foer:

> Monotony collapses time; novelty unfolds it. You can exercise daily and eat healthily and live a long life, while experiencing a short one. If you spend your life sitting in a cubicle and passing papers, one day is bound to blend unmemorably into the next – and disappear. That's why it's so important to change routines regularly, and take vacations to exotic locales, and have as many new experiences as possible that can serve to anchor our memories. Creating new memories stretches out psychological time, and lengthens our perception of our lives.[1]

And here's a fascinating brain fact: every time you learn something, or have a particularly memorable experience, a *new* brain connection is made between two or more brain cells. And the more actively and deliberately we 'exercise' our brains by learning new stuff and having new experiences, the healthier we will be and the fuller our lives will be too.

Alzheimer's Disease International published a report in 2014 which cautiously endorses both physical and mental activity as offering some protection against brain decline. Looking ahead to an ageing Baby Boomer generation, the report suggests more research is required into the potential efficacy of stimulating video games.[2] One sure thing is that the key to keeping our brains sharp for as long as we can is to create new neural paths as often as we can. Autopilot means repeatedly following the same old path and having only a single line of stepping stones across a stream.

You might have heard the following story before, but it bears repeating because it's about what we allow ourselves to miss if we are slaves to our autopilot system and lack a highly active attention to possibilities:

Eyes wide open – a violinist at the subway

A man stood outside a subway station in Washington DC and started to play the violin. It was a cold January morning. He played six pieces by Johann Sebastian Bach, which took forty-five minutes. During the forty-five minutes, since it was rush hour, it was calculated that thousands of people went past him through the station, most of them on their way to work. After the first three minutes, a middle-aged man noticed the musician. He slowed his pace, stopped for a few seconds, and then hurried on to keep to his schedule. A minute later, the violinist received his first dollar tip: a woman threw the money down and continued to walk on without stopping. A few minutes later, a man leaned against the wall to listen to him, but after a moment he looked at his watch and started to walk again. Clearly, he was late for work.

The passers-by who paid the most attention were the children. The first, a three-year-old boy, was hurried along by his mother. He tried to stop to listen to the violinist, but his mother insisted and managed to get him to continue walking, although he turned his head back towards the violinist as they went. This action was repeated by several other children. All the parents, without exception, forced them to move on. (Children, it seems, have a less well-developed autopilot system. They are more spontaneous and live more in the moment from day to day. If they see or hear something they like, they respond to it immediately and fully.)

In the whole forty-five minutes, only six people stopped and stayed for a while. About twenty gave the violinist money but continued to walk on at their normal pace. The violinist collected $32 in total. When he finished playing, silence fell and no one applauded.

How often have you walked purposefully down a familiar busy street, intent on your regular destination, ignoring everything in your path? You probably have many such routes you follow. Routine is what services a near-slavish attention to repetition – hence our dedication to autopilot. In this instance, the violinist ignored by most people was Joshua Bell, one of the finest musicians in the world. He played some of the most intricate pieces ever written, on a violin worth $3.5 million. It was an event organised by the *Washington Post* as part of a social experiment to explore the extent to which we are able to perceive beauty when it is not part of the plan. It's an interesting concept. When something isn't part of the plan, do we take the time to acknowledge it?

Your autopilot brain-rewiring plan

To make a deliberate point of being more conscious of the moment and to experiment with turning the autopilot off (even if it's just for a few minutes).

Brain-rewiring reward

If you occasionally flick your autopilot switch to 'off', you will come across new things and your life will be enriched. You'll see things you've been missing, you'll be in the present moment and appreciate it, and you should have a sense of slowing time down. It might even feel as if there's more of it.

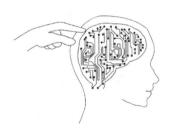

Brain nudges: Four Brain Team tips for countering autopilot

In this section, our Team shares the initial impact of living consciously with autopilot, the thoughts, feelings and subsequent behaviour it prompted, and a few strategies for keeping it in its place.

For some of the Brain Team the idea that they were on autopilot for around 95 per cent of the time was a complete revelation. Even those who were vaguely aware of their autopilot's dominance were spurred into active thought and participation once they had seen the percentage written down in black and white – *Wow, that is a lot on auto. I must be wasting or missing an awful lot of my life.*

Everyone recognised that autopilot played a critical part in their everyday lives, however, and a few expressed anxiety about the risks of mucking about with the switch. After all, isn't autopilot the core coping mechanism for complex modern lives? Isn't it what keeps all the plates spinning?

> *The thought of turning my autopilot off makes me panic as I seem to spend most of my waking hours on multitasking and relying on autopilot, so it will be a big ask of me. With a full-time job, a family and two homes to run, autopilot is my essential aid.*

At the other end of the scale there were a few who rejected the idea of 95 per cent autopilot; they didn't consider themselves to be dominated by routine and felt the idea challenged their self-image and the extent to which they were in control of their lives. The concept of a measly 5 per cent of spontaneous living seemed just plain wrong.

For many people, however, it brought into focus a desire they had been suppressing for years: to slow down, to be more in the present. There was a lot of energy around this thought, holding as it did a promise to counter a bit of the madness of living in such a fast-moving world and to take some time out of it – even if only for a few minutes each day.

Just like our Brain Team, as you become more aware of autopilot's operation in your life, you'll start to see all the routines you rely on and you'll become aware of how many things have become subconscious habits – some good; some bad, perhaps?

I realise how much you feel as though you are looking through a camera, almost blind on both sides and just focusing on what is in front of you. I got a shock at one moment, letting that forward focus go and noticing a picture in the kitchen which had been there for maybe ten years and I had almost no memory of it.

Having read the insight, it made me realise that most of what I do is probably on autopilot: alarm goes off at the same time, leave for school/work at the same time... Work's never on autopilot – too erratic for that – but then the evening routine is the same: school run, dinner, chat and ready for bed by nine!

It made me realise that I do shop in the same place, park in the same space, go to the same café, sit at the same table. Quite shocking to think how boring, but quite pleased to think it's not just me doing this!

1. Become more conscious and aware of how autopilot operates across your life – the process alone is eye-opening!

There are a number of different strategies for playing with the autopilot switch, but the starting point is to identify the subconscious routines and patterns in your life. Then you might decide to use strategies just to be more conscious during that routine, such as stopping on a familiar journey and looking around (who knows when a famous violinist might be playing?).

> *It made me stop and think periodically throughout the day about what I am doing 'right now'.*

> *It made me set the alarm on my phone to ping every quarter of an hour during my working day to remind me to bring myself back into the moment. When colleagues asked me what this was about, I told them I was practising appreciating the here and now. It led to some interesting conversations.*

> *I've stuck Post-it notes on the car and other places, saying 'Wake up!'*

The Brain Team adopted brilliant ways to jolt themselves out of the autopilot setting. Some created mantras, some used electronic or written reminders, and some adopted more extreme measures like putting plastic bands on their wrists which they would periodically *ping*, to snap themselves out of it (literally).

2. Do the same thing but in a different way

You might decide you need more direct ways of jolting yourself into the moment. Try doing something differently, such as taking a slightly different route to work or by walking or cycling instead of taking the bus.

> *I changed my routine for running – both the time and the route. Usually I run having dropped my daughter off at school; this time I got up early (5.30 a.m.!) and took a different route. And when running at this different time of day, I saw a whole load of people I didn't know existed, and it made me think about who they were, what they were doing at that time of day, and it prompted loads*

of thoughts about groups of people I hadn't seen but was now thinking about – lots of positive feelings for all those who had to get up at that time to support the rest of us in services we take for granted.

Instead of going out of the house by the front door and straight down the steps to the car, I went out the back way and round the garden. All our fruit trees are now in various stages of budding/flowering; it was a way to discover spring en route to the office!

I did a little detour on my walk that I had been meaning to do for a while and saw orchids, lambs and ducklings.

You see? It really doesn't have to be a mighty change.

3. Identify and tackle routine creeps

You might identify actions which have become routine – these are the routine *creeps*: small, apparently insignificant things that you find yourself doing without really pondering the whys and wherefores; actions that you might want to take a break from, or at least pause or postpone occasionally, like the habitual glass of wine when you get home that you pour without thought, or the compulsion to eat the last biscuit in the pack, or to eat automatically.

There are two reasons for eating automatically. One concerns our ability to chew, swallow and digest without conscious thought. The other relates to eating without thinking about the food itself and then eating too quickly – going from full plate to empty plate in minutes, without savouring the food or the experience. For the Brain Team, focusing more on the food they were eating resulted in them enjoying food more, eating *less* and, much to their delight, losing weight.

I tend to eat on autopilot, and usually without giving thought to all the things you get from food, particularly taste, unless it's special or spectacular.

I realise that I have a very bad habit of going into automatic when eating on my own – just throwing food in, as if it came from nowhere and was of no concern to me. I am really going to try to do something about that – at least to eat more slowly and to think about where this precious stuff (which I take so much for granted) has come from. Also, by thinking about eating, I have lost 2 lb! (Most of the time, when I think first, I decide I don't really want whatever it is.) I've made no other changes to my diet!

What is interesting is, even when this exercise is fresh in your mind, how easy it is to revert to 'autopilot', particularly when you're busy at work or tired at home. I had to make an effort not to turn on the TV or pour a glass of wine when I got back from work and listened to music in the garden instead.

4. Take a deliberate timeout and do something that's not part of the plan

Whether it's putting down your phone, eating lunch away from your desk or letting the washing machine go unloaded for a change, it's a breath of fresh air and can be a moment of personal liberation.

I took a proper break at lunchtime away from my desk each day and spent time either with work colleagues or by myself – simple stuff, but I haven't taken a lunch break for years. Quite liberating and the key to it was getting away from my desk.

I'm going to start with spending more time just sitting in my garden rather than working in it. Maybe I can learn to forgive the weeds and see them as simply flowers that have found their own way there.

Rather than waiting until the weekend to see an exhibition, I made an effort to leave work early (unheard of) and popped in to see it during the week. It felt as though I had been freed of the constraints of work: a little escape from my normal routine of staying late, which restricts any such opportunities.

In the past, returning from anything – a day in London, a weekend away, a holiday – has been horrendous for me and more especially for my immediate family. I rant, I become manic, I put clothes straight into the washing machine, put suitcases in the loft, open all letters, answer all letters and then feel thoroughly disgruntled, and all holiday spirit has well and truly gone. I am now definitely going to approach the return, the touchdown, with much more simplicity. Most things can wait, while I sit down and chill.

I realised that checking and holding my phone 24/7 (even sometimes taking it to bed) had become the norm, but today I took a day off the grid: I turned my 'must check' autopilot off and have not looked at my emails for a whole day or answered my phone. Instead I spent (much more enjoyable) time with my children.

Reflections from the Brain Team: What happens when we're more conscious of our autopilot

'Taking back control' is a phrase liable to strike doom in the heart of anyone who's lived in the 'Brexit years'. But one of the significant discoveries among the Brain Team was that although autopilot helps you *to survive,* it does not help you *to experience life fully* – while having control of the switch does! As people became more thoughtful about the impact of autopilot on their lives and, in particular, when they were consciously stopping themselves from auto-led behaviour, or consciously holding off turning the autopilot back on, they developed a real sense of satisfaction. It made those times feel richer.

> *I feel my house is rather out of order, but it is 10 p.m. on Sunday night and my week has been so memorable and happy.*

The power of stolen moments – when you consciously turn autopilot off or refuse to switch it back on – was captured in the words of one of the Brain Team who said, 'It was about giving yourself the gift.' It's a great idea. And what is glorious is that there is no end to the gifts (*free* gifts, after all) you can give yourself. You can give yourself the gift of not feeling you have to rush, the gift of taking your time, the gift of just standing for a moment or of giving yourself permission to do things differently. And when you are not doing things automatically – when you are, perhaps, doing new things or going about old things in a new way – you are creating more stepping stones across the stream and thereby generating more neural pathways for your brain.

The beauty of turning the autopilot off is that even the simplest, smallest things can make a memorable impact. Here's a perfect example. One of the Team was walking her dog one day and was in a hurry to get home. She caught up with another dog walker, an elderly man with a pack of unfeasibly aggressive small terriers. Normally they automatically gave each other a wide berth, but this morning – because she was working on switching off her autopilot – she stopped and said hello and, as they stood briefly to chat, the man pointed out a rare orchid blooming in the meadow grass beside them. It was one of those moments – a moment in which, if she had not been consciously turning off her autopilot, she would have been deprived of seeing the beauty of a rare flower in a field and a moment of rich connection to boot. We can all try to take tiny steps like these, one at a time, and see how little changes help to build one's confidence, enabling us to flick the switch to 'off', knowing we can switch it back on just as fast.

So, whether it's looking up, down or sideways while walking in an urban street, pausing for ten seconds on a regular journey or making tiny changes in a routine like the Brain Team did, it can take only seconds to add a whole new sense of colour to your life.

These benefits are best captured in the reflective comments of the Brain Team:

Coming off autopilot is making me thinner, holier, braver, calmer. I'm generally feeling more in-control. What's not to like about that?

I will definitely stop and think before I start doing the same routines day in, day out. Putting my autopilot on hold can be entertaining, informative, enriching, stress-relieving... provided one remembers to turn it off – and back on when needed.

Being more aware of each moment has allowed me to realise what it is that I really value in my day, what is positive and what is negative. I have endeavoured to push my boundaries in tiny ways, but this has made me feel much more confident and at ease with myself.

Each small step has built up my confidence. I feel a certain inner self-assurance. I am thinking as I go, rather than contemplating the next thing that might arrive. Various apparently meaningless social interactions, when tackled with active thought, help to motivate me and help me to grow. Even the smallest positive events contribute to this, and I am increasingly seeing value in the smallest actions. It will certainly contribute positively to my personal happiness, self-esteem and my relationships.

One of the Brain Team's biggest realisations was that even very small changes could have an impact – and that is the mantra you will keep hearing: small changes can deliver big rewards.

I think that being aware of how you can just alter the smallest thing, and suddenly all your senses are slightly heightened, will be a useful thing to remember. Small tweaks in my daily routines could alter the course of my life and introduce me to new ideas, people and places.

Reflecting back on the autopilot journey included moments of shock (*'95 per cent autopilot? You're kidding!'*) and fear (*'I can't live without it!'*),quickly overtaken by an emerging feeling of excitement about flicking the switch off and on. The overriding takeaway for the Team was, yes, of course we need the autopilot in many aspects of our lives, but it's good to be more conscious of its pervasive presence, to be able to check now and again when it is better switched on and when we can be confident about turning it off – and that when we turn it off, there's a new world to take notice of. The Brain Team were liberated and gained a sense of adventure from taking charge of their autopilot setting. It was fun to do, it was easy to do, and good things resulted.

I need to continue to access the sense of freedom I get from realising it doesn't have to be that way. Very liberating.

Generally, thinking about the concept of autopilot made me put more down that irritates me, pick more up that energises me and spend more time with the people I care most about. Going with the flow of the day, instead of sticking to my routines, gave space for other, unexpected and delightful things to occur.

The words 'brain insight' have been imbued with thought and taking time, so it's become a sort of mantra, and whenever it pops into my mind it makes me sit there a little longer, think a little more, look at the view a little longer.

Brain-rewiring awareness: Autopilot

Many of us struggle with the demands of our busy lives and rely on our autopilot to survive. But what we have seen in this research is how increasing self-awareness of the autopilot and acknowledging how it operates in our lives can be deeply empowering. Awareness alone can be shocking (when we see how much of our daily lives are run on autopilot) but also hugely enlightening. The two-stage process of awareness and action – that is, a growing conscious understanding that can lead to more conscious action – allows us to explore a number of strategies to play with the autopilot switch.

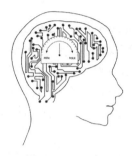

Summary of four brain nudges to counter autopilot

1. Become more conscious and aware of how autopilot operates across your life – the process alone is eye-opening.
2. Deepen your consciousness of a regular routine and see new things just by being more in the moment.
3. Engineer slight changes to your routines to jolt yourself into more active consciousness.
4. Take a timeout and do something that's not part of the plan.

And just remember: it's all about taking back a little control.

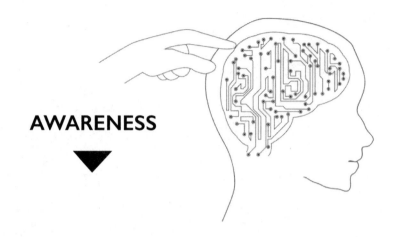

AWARENESS

▼

UNDERSTANDING

▼

ACTION

▼

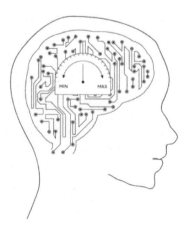

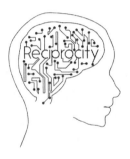

2. The Reciprocity Bias

We're wired to respond in kind: How to trigger positive ripples and prevent negative backlash

Reciprocity bias brain-wiring insight

Our brains are wired with the tendency to behave towards others as they behave to us – this is called the reciprocity bias. In the words of the very wise:

> There is one word which may serve as a rule of practice for all one's life: reciprocity.
>
> Confucius

> If you want others to be happy, practise compassion. If you want to be happy, practise compassion.
>
> Dalai Lama

We gave our Brain Team this reciprocity insight to work on over the course of a week, inviting them to go out of their way to act generously and engender positive mindsets as far as they could:

> Our brains are wired with the tendency to behave towards other people in the same way that they behave to us. This is called the reciprocity bias. We think of it almost purely as a positive

impulse, but it can work both ways. It is equally powerful used positively or negatively. The other really interesting fact is that acting generously or positively creates immediate activity in the area of the brain that is stimulated when you receive a reward yourself. It's a win-win – doing good to others makes you feel good too. And get this: every positive act you carry out, however small, can generate a ripple effect as you encourage the people around you to do the same.

How reciprocity bias operates in our lives

In research by Professors James Rilling and Gregory Berns, neuroscientists at Atlanta's Emory University, participants' brain activity was recorded while they were given the chance to help someone else.[3] When they did so, this triggered activity in portions of the brain that are activated when people receive rewards or experience pleasure. And every positive act you carry out, however small, can generate a ripple effect as you stimulate the reciprocity bias in the next person.

> It is one of the most beautiful compensations of life that no man can sincerely try to help another without helping himself.
>
> Ralph Waldo Emerson

Negative reciprocity – the flip side of the reciprocity coin – is also a heavy hitter. Some psychologists sometimes refer to negativity as 'punishment' and believe that we are more likely to punish 'bad' behaviour than to reward 'good' behaviour.[4] We can see this tendency hard at work in our own lives. A driver undertakes us in an aggressive manner and we're likely to drive faster behind them; a shop assistant is curt and unfriendly, so we don't say thank you or smile; we send someone a gift and they don't write to thank us, so we vow not to send such a nice gift next time. And, of course, there are far worse examples of brutal retaliatory actions taken in conflict situations.

Bringing positive reciprocity bias into your everyday life

There's a fair bit of inspiration out there (we've already touched on Marks & Spencer and their adoption of reciprocity bias at the checkout – their version of the 'Have a nice day' effect). A great example of reciprocity bias in action was Acts of Kindness, a project that was part of the Art on the Underground programme run by Transport for London during 2011–12. The artist Michael Landy asked people to post their experiences of the kindness of strangers on the website, where they are listed one after the other in a cumulative embrace of goodwill.[5] Many of the stories demonstrate some glorious reciprocal rippling. Here are some examples:

> Coming home after a Xmas party on the Tube we got talking to a lovely man who was carrying boxes of crackers which after a while he started to share with all the people in the carriage – all strangers, but we were all putting on hats, taking turns to read out the jokes and sharing the gifts ... It was just lovely and brilliant fun. The best Tube journey, and if the man should be reading – THANK YOU!

> Leaving Oxford Circus, into torrential rain. As I stare at the rivers forming in the gutter, a lass walks past with a golf umbrella. 'Plenty of room under here for two!' she said. And there was!

> I saw a lady carrying some heavy suitcases near King's Cross, so I offered her a hand up the stairs. She accepted but then we both began to struggle and a man came up and also offered

help. He took one suitcase while I took the other. Another man saw this and offered to carry my bag. We ended up all carrying each other's bags and laughing all the way up the stairs. We were all smiles.

Your reciprocity bias brain-rewiring plan

To use this wiring more purposefully to create positive waves or to stop negative waves.

Brain-rewiring reward

This is a simple, recognisable piece of brain wiring. It's also known as 'Practise what you preach', 'Do as you would be done by', 'One good turn deserves another', 'What goes around comes around', 'If you want a friend, be a friend', and so on and so on. What we're saying is not news to you. The reciprocity bias is, perhaps, commonplace. And we guess sometimes this means that we don't give it much headspace, don't really credit it with having much of an impact on our lives. In fact, the opposite is true. If you work through the familiarity and pull the concept to the front of your mind, you will see how much power it has:

- ⊙ It can trigger positive ripples
- ⊙ It can confront negative energies
- ⊙ It can prevent negative retaliations
- ⊙ It can give you breathing space
- ⊙ It can enrich relationships
- ⊙ It can make the world a better place

Maya Angelou said: 'I've learned that people will forget what you said, people will forget what you did, but people will never forget how you made them feel.' And that's worth bearing in mind as you reflect on the reciprocity bias.

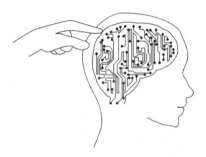

Brain nudges: Four Brain Team tips for embracing the reciprocity bias

This section shares the initial impact of living consciously with reciprocity bias, the thoughts, feelings and subsequent behaviour it prompted among the Brain Team, and a few strategies for applying it most effectively.

1. 'To give is to receive' – keep this mantra at the front of your mind

It might sound a bit preachy, but that simple statement packs a punch. Think about it in everyday life. When can you identify with it or see it around you? It's contained in the smallest of moments: smiling at a stranger, letting a driver out of a junction ahead of you, saying something complimentary to a friend. When you commit acts of positivity, you'll feel good – and things will be better for others too.

> *I was thinking about the brain insight sitting on my train and was pleased I had managed to get a seat as it was so busy. A guard got onto the train with a young mother and her baby in a pram, and looked my way. He asked me if I would mind giving up my seat for them because there was plenty of room for the pram where I was. I said of course, and smiled, even though I knew I was unlikely to find another seat as the train was full. I stood in between the carriages and after a while the guard came back and asked me to come with him. He said, 'I want to put you in first*

class for being helpful.' I was thrilled. He said, 'It's one of the few things I can give.' Then he turned to another member of the train staff and said, 'Give this man a free drink!' When I left the train, I told the guard how kind he had been and if only there were more people like him in the world. It was a bit cheesy, I know, but he smiled a huge smile and I thought to myself, I wonder where this positive wave will end...?

In his book *Influence: The Psychology of Persuasion*, Robert Cialdini says that reciprocity is a powerful drive found in all societies, and can often turn a 'no' into a 'yes', even for requests that would normally be refused.[6] Sometimes when you're walking towards someone, you find you've made a negative decision about the kind of person they are; you might judge them to be grumpy or mean-spirited in some way, even though they've done nothing to deserve it (except not to look cheery). Try flashing them a big grin as you pass them. Then, when they smile back, you'll have changed your own negative into a positive, and given them something to smile about into the bargain. Have in your head the image of the character George Bailey in director Frank Capra's movie *It's a Wonderful Life,** and imagine the world a poorer place for your not having been in it.

And don't forget, the flip side of the bias is that acting negatively can work reciprocally as well – what goes around comes around... It doesn't matter which way you work it. Remember the urge to punish is strong in all of us, so watch out for careless negative actions, even in the simplest things – the unintentionally curt tone of an email, for example – which can result in negative reciprocity.

---

* In *It's a Wonderful Life* (1946), George Bailey wishes he had never been born and is given a glimpse of what life would have been like if he hadn't.

2. Conceptualise the ripple effect spreading outwards

Think about the earlier idea of random acts of kindness – a chaos theory of kindness, a tidal wave of warmth. Start one today – who knows where it might end?

As you do something good for someone else, it can become more powerful if you have a mental image of the ripple effect of what you have done. Some Brain Team members also talked about a way of conceptualising positive actions in terms of 'paying it forward', i.e. what you give to others you hope they will pass on rather than give back. It means the waves have a greater reach. It's a weirdly energising and wonderful thought – that we might have the power to make someone, somewhere, do something decent or kind, or even lifesaving, just by smiling at one person (the first link in the chain, if you like) and making them feel good. Before you know where you are, drivers are smiling and making way for each other all over the world (well, we can dream).

> *Because maybe the person we smiled at – who did not seem to respond to us – fifteen minutes later smiled at a person who had until that point thought this person hated them, who was then so overjoyed they gave everyone they met a compliment, etc., etc.? The waves will keep rolling and rolling.*

> *A bit like the butterfly flapping its wings and causing a hurricane on the other side of the world, hopefully one of my positive ripples could result in an extreme act of kindness somewhere in the universe! From my own perspective, a lifetime of positive actions will keep me happy, healthy and in control of my life.*

Here are a few easy ways you can add a bit of reciprocity into your life, too. Do one of the following things *today* to achieve the same effect – and it will make you feel better at the same time:

- Be friendly to everyone you meet.
- At work, the school gates, the supermarket or the gym, say something complimentary to everyone you spend time with – 'Love that colour on you!', 'You look fabulous today', 'How lovely to see you!', 'You've made my day!' etc.
- If you can't quite bring yourself to say a compliment out loud, make one *internalised* (silent) positive comment about people you pass in the street (*Lovely shoes... Great hair... What a good dad... You get my drift*).
- Smile and say good morning to a stranger.
- Hold open a door for someone.
- Ask a friend how they are and really pay attention to their answer.
- Let someone go ahead of you in a line.
- Let another driver out of a side road or give way to them.

> *People have given me so many lovely personal compliments this week – about each of the children, about my appearance, about the courses I've run, my cooking, smiles in the street when I'm out running – and each of these definitely felt like a warm hug and a sense that all is well with the world, whatever my anxieties.*

It's quite likely of course that your cheery greetings and smiles to strangers will risk causing confusion or concern among your fellow humans, who may be more used to less friendly folk, but you shouldn't let that stop you. A couple of Brain Team members had this to say about some of the responses they provoked:

> *On my morning run I decided to smile at all the stern-faced walkers and runners, and on the first day got a range of responses, from joy to 'Oh no, he's mad!' But by the second day people almost saw me coming and I got so many more smiles from the regulars, and I bet they had a better start to their day too!*

I had mixed reactions to my friendliness and hellos to strangers. One guy in the Tube station looked so startled and nonplussed, I felt like a real idiot! In general, I think people in London are very, very wary of contact with others, and their natural inclination is to automatically think someone's a bit soft in the head if they look remotely friendly – which is, of course, very sad.

3. You can use reciprocity to stop a negative wave as well as starting a positive one (it's a double whammy), so try reciprocity as a pre-emptive strike

It's your choice. You can get mad at the guy who refuses to give way to you at the roundabout or you can focus on stopping (and certainly not starting) a negative wave. You can get all riled up with fury (a feeling that's likely to stay with you the rest of the day) or you can just let it go, take a breath and get on with the business in hand.

Here are some other ways of derailing the negatives:

- ☉ If someone doesn't let you in at a road junction, make sure, the next chance you get, that *you* let someone in – whatever you do, don't take your frustration out on everyone else.
- ☉ If someone says something that sounds a little unkind about someone else, say something good about that person to deflect the negative.
- ☉ If you are faced with an angry co-worker or sales assistant, see if you can apply the positive reciprocity bias to turn things around. Be bold: to a co-worker offer a cup of coffee; to a sales assistant some genuine-sounding sympathy: 'Tough day?'
- ☉ If you get an email or text message from someone and it seems to have a negative tone, try to reread it from a different angle – reframe it – until it sounds more positive. You'll prevent a prickly, negative reaction from kicking in and your fingers from bashing out an angry response.

I became very aware of just how much negative energy I give off. I caught myself thinking everything from mean thoughts about people and 'poor little me' thoughts about my business, to little internal criticisms of what strangers in the street were wearing, how they had their hair and what their faces looked like! I had to very consciously remember to think positive and be gracious.

I just said a smiley good morning to a very grumpy traffic warden. Not much reaction, but I felt pleased with myself! Hope he is cheered up inside even if he doesn't show it!

I thought a good thought about everyone I saw – walking past them, on the train, in shops – I like their hair colour or That skirt looks good or Nice nails. Especially of people I normally would not notice or would be critical of in some way.

Several members of the Brain Team discovered the power of purposeful application of the reciprocity bias to change a negative into a positive, or to disarm a downward spiralling situation. It works.

I've tried to diffuse situations with children when crossness threatens to get out of control on both sides. It's sometimes so hard to do – it takes real effort – but it always, always works.

While I was waiting in the school playground, I pushed myself to talk to some of the mothers I least like! I don't know if this gave them a warm feeling, but they all reciprocated in a friendly manner. I was glad that I did it as, rather than feeling uncomfortable in the playground, it meant that I had turned it into a positive experience for me, which in turn seemed to be positively received.

There have been a few family incidents where a situation would normally have escalated but I was more conscious of the effect my behaviour would have on everyone and therefore everything stayed calmer and more relaxed, and this was definitely a better way of dealing with the situation. Family tension was definitely defused a lot more this week because of the reciprocity!

The fact that we can potentially get ourselves into a good and positive mindset prior to, say, a meeting or turning up for work, by simply being 'nice' to people beforehand has practical benefits from a business and social point of view.

One of the deeper reflections that came out of this brain insight was the power of non-verbal communication. It is amazing how easy it is to give out negative vibes without being aware that you're doing it. Just think about the following non-verbal cues for a few moments – they're all negative:

- ⊙ lack of eye contact (suggests inattention, shiftiness, untruthfulness)
- ⊙ leaning back in your chair (superiority, boredom, lack of interest)
- ⊙ fidgeting (inattention, nervousness, boredom)
- ⊙ crossing your arms (hostility, defensiveness, aggression)
- ⊙ checking your phone (the ultimate insult)

4. Listen more carefully in conversation with others

Reciprocity can take lots of creative and subtle forms. Concentrating fully on what someone is saying can trigger deeper conversations and make for better relationships. The beauty of how our brains are wired is that engagement tends to trigger engagement.

It's true that to make conscious a positive and open 'attitude' to someone that you are with does create of flow of positive energy... like showing a beam of light that bounces back and illuminates you as well as the dark place you're in. I certainly consciously listened to a friend, reflecting back acceptance and interest, sending out positive waves, and it made me really connect with her and really get involved, instead of sitting there, itching to get on to the next task in my day. The more animated and conscious I was, the more I felt it returned.

Reflections from the Brain Team: Living consciously with reciprocity bias

This is such a rewarding insight to explore because it nearly always works, and results are pretty much instantaneous. The Brain Team loved putting the simplest ideas into action. What is clear is that, when you think deliberately about how you're acting and reacting to things and whether you're generating positive or negative vibes, it makes a real difference. People will notice what you do.

Reciprocity bias is a perfect example of how a small change from the subconscious default setting to active, focused practice can make a big difference. Just thinking about it creates a greater sense of the power and, critically, the opportunity and motivation, to use it in everyday life.

These are the Brain Team's *big* takeaways:

A simple reciprocal smile can improve relationships

I did have a breakthrough with a totally miserable neighbour who has consistently ignored us. I decided enough was enough, and I smiled at her and said hello. She didn't reply but half-smiled. However, it seems to have worked as the next day she actually greeted me! What a turn up – so much nicer to have amicable relationships with neighbours!

As a head teacher, I have a number of parents who have great difficulty in acknowledging the greetings I give, but I was determined that I would be greeting everyone regardless of the response I did or didn't get. By the end of the week, one parent who has been ignoring me for quite some time (because she wasn't happy with a decision I made) actually gave me a 'Morning', albeit rather grudgingly and without looking at me. It's a good start… and I'll keep working on it.

I have probably been labelled as the psycho of the towpath after Wednesday afternoon. I cycled, not only smiling at everyone I met, but greeting them as well: 'Hi', 'How are you?' and even 'Sorry you had to leap out of the way!' to one particular chap who came upon me all of a sudden round a bend. To a man, they responded; all of them smiled and most said something pleasant. All in all, a thoroughly exhilarating exercise which felt hugely positive for me – and I think it was for my 'greetees' too. A negative ripple was prevented because I do usually use my bell to warn that I'm coming: for some reason, bike bells tend to irritate people in the same fashion as crying 'Get out of my way! I'm coming through!' might.

It's so interesting to think what a smile or greeting does for one's psyche – it just sets you up for the rest of the day.

I will use this insight to make me think about my actions and what effect they may have on others. I will try and hold my tongue, and think of the effect of something before I speak.

And make no mistake: those positive ripples will come back to you.

Reciprocity bias can stop negative waves in their tracks

It is so easy to respond to a negative with a negative. Someone looks at us in what we perceive as an unfriendly way, and it makes us feel unfriendly too. Someone goes on the attack, and you attack back. Both scenarios can end in a not-very-happy place – for everyone involved. But if we think creatively about the reciprocity bias then it is amazing how little changes can make a huge difference. When I occasionally lose my temper at home, my wife makes me laugh and it quickly diffuses the situation. The Brain Team found that their week living with the insight gave them ideas for how to deal with a whole range of things. And the simplest ideas are often the easiest and the best. It's easy – you can stop something negative from happening by doing something positive. If you smile at someone, it's difficult for them to stay furious; and neither of you will feel worse for the exchange – you can only feel better, and more connected. Small change, big difference.

The understanding of how negative ripples can keep building, along with the idea of using positive waves to counter negative ones, is deeply empowering and will become part of many of the Brain Team's life-armoury in the future.

In applying this, I was able to keep the monster calmed on several occasions when I wanted to say something negative, or complain or get angry. By the end of the week I had a much better outlook on life as I felt my moods were elevated by all the positive experiences of the week, and the lack of negative ones.

I can see this will help me to diffuse an argument with a positive comment or try to find a way of seeing the argument from someone else's perspective. I'll try to be less stubborn, too. I could see myself using the reciprocity bias to bring out a potentially much more positive result in business and my everyday life.

I think if you are aware of negativity and take action very quickly – i.e. nip it in the bud – it could prevent huge repercussions and therefore have a big impact. And, therefore, I suppose small positive actions can lead to massive changes and achievements, which gives hope to the world.

It's probably more effective to act without expectation

Some team members thought deeply about whether the motivation for positive acts could be either altruistic or self-serving – and there was a distinct acknowledgement that when other people failed to respond gratefully to, say, cheery greetings or selfless driving, it gave rise to annoyance and irritation. They had made conscious efforts to check this – otherwise, it dawned on them, the reciprocity bias would be dead in the water or, worse, likely to trigger a negative wave. Application of the reciprocity bias works best if you try really hard not to expect anything in return; you have to adopt a biblically philosophical approach and assume that while some of your good vibrations may fall on fallow ground, most will take root.

Much of it I found was not necessarily in the actions themselves but in the mindset – yes, do kind things for people but do so graciously and the kindness gets passed on. If you are kind for kindness' sake and not for the thanks that you may or may not receive, kindness can become more of a habit. If you wait for the thank you and then don't receive it from those who may not have noticed, or may be selfish or whatever, the negativity and self-centredness creep back in.

We can of course remind ourselves that our brains *are* getting a reward for our good deeds – whether we realise it or not!

Even in our short experiment, the Brain Team began to see that what goes around comes around

I've been practising bigger smiles and more conscious eye contact while greeting people as well as driving less aggressively! It obviously worked, as the workman I had greeted on the way downhill one morning saw me on my way home; his face lit up and he seemed absolutely delighted to see me again and wished me a very good day!

I noticed an interesting development that, in trying to drive in such a way as to be courteous and to let out other drivers etc., it appeared that non-aggressive driving caused far more people to let me out or to give way to me! Either courtesy creates more courtesy or all the other drivers were part of the Brain Team! It was quite noticeable and made for much more pleasant driving.

The trick is to keep the concept of reciprocity front of mind

It might help to examine the necessary steps to achieve this:

⊙ How do you make it become part of you?
⊙ How do you sustain it?
⊙ How do you keep doing it when you're in a hurry or under pressure?
⊙ What do you do when it's not reciprocated?

Some of the team thought they should start each day asking themselves three questions:

⊙ How can I make someone's day today?
⊙ How can I create a positive ripple?
⊙ How can I stop a negative one?

Others wrote themselves directives or daily behavioural nudges, which included:

Take a second to smile and say hello to colleagues when they come into the office. And begin as many interactions as possible with a smile...

Ensure that at least once each day I'll find something to compliment a stranger on (I like your hair/bag/tie).

Look for something positive to say about, and to, people, even where the interaction is a difficult one. Or especially then.

Greet anyone I meet with a genuine-sounding 'It's nice to see you. How are things?' Take an interest, be positive and constructive.

Some even used technology for a daily nudge:

I plan to apply this insight more religiously. I've changed my password at work, encoding into my new one a reminder of this insight so that I do not forget to apply it. Typing the new password all week should be a regular reminder to me to be cautious not to cause bad ripples.

Brain-rewiring awareness: The Reciprocity Bias

Why not conceptualise the reciprocity bias as a brain tool which has the power to change your environment? Actually, why pussyfoot around? Why not conceptualise the reciprocity bias as a brain tool which has the power *to change the world*?

If we all set out more deliberately to radiate *positive* vibes, and we imagine the butterfly wings of our positive intentions triggering individual positive experiences for one person after another and all that positive action spreading out in ever-widening ripples (with our initial trigger at the centre), and if we act consciously not to generate negative vibes, or actively to nip them in the bud, there's no knowing what we could achieve. Having one's eyes wide open in terms of one's own behaviour and its potential impact on others is pretty awe-inspiring.

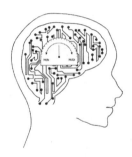

Summary of four brain nudges to embrace reciprocity bias

1. 'To give is to receive' – keep this mantra at the front of your mind.
2. Conceptualise the ripple effect spreading outwards. Who knows where it might end? Think about the idea of random acts of kindness – a chaos theory of kindness, a tidal wave of warmth. Start a ripple effect today.
3. Use reciprocity to stop a negative wave.
4. Listen more carefully in conversation with others.

Go on – start triggering those ripples!

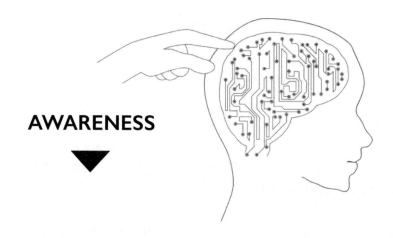

AWARENESS

▼

UNDERSTANDING

▼

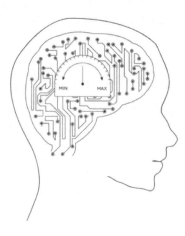

ACTION

▼

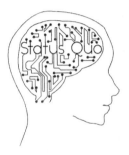

3. The Status Quo Bias

We're wired to keep things the same: Learn how to rock the boat

Status quo bias brain-wiring insight

By and large our brains are wired with the tendency to maintain the status quo and to avoid change. It's because we like our lives to be dependably predictable and don't much like putting our heads over the parapet. We gave our Brain Team this status quo insight to think about over the course of a week, inviting them to say yes instead of no, and try other ways of finding adventure in the ordinary:

> You may recognise this little truth: that we don't really like change and have both a conscious and subconscious tendency to try to keep things just the way they are. This is called the status quo bias. What this means is that we are predisposed to reject ideas, actions and suggestions that would cause change (however big or small) and disrupt the familiar in our daily lives.
>
> We are wired with a tendency not to rock the boat and to put off doing things which are just not part of our everyday plan – we want to keep our life running in a straight line. This is why you might want to say no to things more often than you want to say yes.

But shaking up the status quo and pushing ourselves outside our comfort zones keeps us a little sharper.

How the status quo bias operates in our lives

We're wired to prefer the familiar, to stick with what we know; we can't help ourselves.

The following excerpt is taken from the 1953 novel *Fahrenheit 451* by Ray Bradbury. The story depicts a dystopian society where, to prevent citizens from knowing, learning and wanting more, books are not just banned, but burned. Guy Montag is a fireman who discovers a new world when he meets Clarisse, a rebellious, boat-rocking young woman who asks questions, loves nature and sparks in him a zest for doing away with the status quo. She introduces him to a character called Granger, who, together with a group of fugitives, has taken shelter in the woods to escape the regime and to read and memorise books. Here Granger quotes his grandfather's inspirational words:

'I hate a Roman named Status Quo!' he said to me. 'Stuff your eyes with wonder,' he said, 'live as if you'd drop dead in ten seconds. See the world. It's more fantastic than any dream made or paid for in factories. Ask no guarantees, ask for no security, there never was such an animal. And if there were, it would be related to the great sloth which hangs upside down in a tree all day every day, sleeping its life away. To hell with that,' he said, 'shake the tree and knock the great sloth down on his ass.'[7]

Without the tree shakers – those who have the courage to knock that sloth on his ass and go their own way – our world would be infinitely poorer. In fact, many of the world's most successful entrepreneurs and creators are the ones who probably started out by rejecting the status quo. An advertising campaign made for Apple in 1997 by the Los Angeles office of TBWA\Chiat\Day was called *Think Different*. It is a magnificent salute to those who have gone their own way regardless of the herd, and a great call to action for the rest of us:

> Here's to the crazy ones. The misfits. The rebels. The troublemakers. The round pegs in the square holes. The ones who see things differently. They're not fond of rules. And they have no respect for the status quo. You can quote them, disagree with them, glorify or vilify them. About the only thing you can't do is ignore them. Because they change things. They push the human race forward. And while some may see them as the crazy ones, we see genius. Because the people who are crazy enough to think they can change the world... are the ones who do.[8]

It would be easy to regard the status quo bias and autopilot as the same, but there are some important distinctions. Autopilot is about our lives being a series of predominantly subconscious routines and about us trying to find ways to be more conscious of these moments; to wake up, open our eyes to the journeys and not let everything race past us. Status quo bias relates to our natural tendency to want to keep things the same. The aim here is to change things we do, to add new things, to open up to new experiences and, in so doing, to re-energize ourselves and our lives – to rock the status quo!

Let's be honest, though: despite the glorious-sounding rhetoric of the Apple ad, most of us don't really like that much change; we are comfortable with our routines and with the permanence and consistency around our lives. What's not to love about a familiar routine and knowing where we stand? When we asked our Brain Team to challenge their status quo-based responses to the routine events of their lives, it prompted anxiety. A lot of this stuff has taken years to set up and form around us, and there is no way we are going to get rid of it just like that. The simple truth is that we tend to try to keep things the way they are. Here's an excerpt from a paper published in 1988 in which the term status quo bias was coined by its authors William Samuelson and Richard Zeckhauser. They explain that this an incredibly common trait, regardless of circumstances:

> Most real decisions, unlike those of economic texts, have a status quo alternative – that is, doing nothing or maintaining one's current or previous decision. A series of decision-making experiments shows that individuals disproportionately stick with the status quo ... Faced with new options, decision-makers often stick with the status quo alternative, for example, to follow customary company policy, to elect an incumbent to still another term in office, to purchase the same product brands, or to stay in the same job.[9]

So, it looks like we are wired with a tendency *not to rock the boat* and to put off doing things which are not part of our everyday plan because we prefer our lives to run in a straight line. The result is our brains can get stuck on status quo as a default setting. And while I'm not suggesting that sticking to the status quo on an everyday basis will bring about disaster, it *might* result in your being stuck in a rut; in a sort of action and opportunity paralysis. It will mean you miss out on stuff. It's an 'If you do nothing, nothing will happen' scenario. But going against it can be hard work.

Studies in the US show that one of the negative impacts of status quo bias emerges in the context of medical non-compliance – where patients suffering from, say, heart disease or diabetes fail to take medication or to initiate contact with doctors.[10] Medical non-compliance is estimated to increase healthcare costs by $100 billion per year in the US, so it's worth wondering if anything can be done to tackle it. Sometimes patients stick with a decision already made. We've probably all done this. Household insurance renewals, anyone? But medical non-compliance can mean the difference between having a heart attack and *not* having one, so how to change that behaviour? One way is to work harder to encourage compliant behaviour from the outset, getting patients to try the recommended behaviour once, so that compliance is more likely to become the default setting from which the patient does not deflect. Status quo is a powerful beast.

According to research by UCL scientists that examined the neural pathways involved in status quo bias in the human brain, the more difficult the decision we face, the more likely we are not to act.[11] It can be a case of 'when in doubt, do nothing'. As part of the research, sixteen participants were asked to act as tennis line judges, deciding whether the ball was in or out. They had to look at a cross between two tramlines on a tennis court on a computer screen while holding down a 'default' key, and each time the ball landed the computer signalled which was the current default option – 'in' or 'out'. The participants continued to hold down the key to *accept* the default but had to release it and change to another key to *reject* the default. Results showed a consistent bias towards the default decision (the status quo, in effect) with participants 'agreeing' with the computer decision, which led to errors. As the task became more difficult, the bias became even more pronounced.

In another, slightly more gruesome-sounding experiment, participants were recruited to undergo electric shocks.[12] The 'cover' story for the experiment shared with participants was that it was to determine the link between participants' subjective anxiety and their physiological responses. The real point of the exercise was to ascertain the extent to which participants would stay with the default option, the given status quo, even when it was clearly less appealing. Specifically, participants were given the option of reducing their anxiety while waiting for an electric shock by opting to get it over and done with quickly, rather than waiting for it to happen at an unknown time. Participants were randomly assigned to one of two groups. One group had a default option – that of not reducing the waiting time – while for the other group there was no default and participants had to press one of two buttons: one to reduce the waiting time, the other to keep the waiting time unchanged. In the default-option group, participants had just one button, which they could press to have the waiting time reduced; or they do nothing and the waiting time would remain the same. It was predicted through pre-experiment surveys that the majority of people would want to reduce the waiting time between starting the trial and receiving the shock, as a longer wait is associated with heightened anxiety – especially as all participants had been told that the shock level had been calibrated such that it would be at the 'maximum level' they could tolerate.

What actually happened? In the no-default group, who had to press one of two buttons, participants opted to shorten the wait around 75 per cent of the time, whereas those with the default option – press one button or *do nothing* – chose to reduce the wait time only around 40 per cent of the time. It's a statistically significant finding and clearly reveals how status quo bias doesn't always operate in our best interests.

We are also wired with a tendency to conform to the status quo of others around us. We don't like to stand out and be different; none of us want to be the round peg in a square hole. And, I guess, on an evolutionary footing, this is probably a piece of brain wiring that has been key to the survival of man. It's a tendency that is hardwired into us and is part of our survival wiring; standing out, or not doing what others did, was probably a quick recipe for early extinction. These days, however, it has evolved into a kind of peer pressure and is founded on the concept of us wanting to do what others are doing, and not wanting to upset those around us. It's why we so often trust in the whim of the majority and blindly follow the herd. And we can't help but feel the eyes of other people upon us and sense their judgement if we don't conform. The status quo bias nudges us to want to do what we believe most other people are doing; it's what makes us feel comfortable. And as the huge growth of social media has increased the power of the herd, and more and more aspects of our lives and our communication are public and on display, this impulse to conform can only intensify. We are all subject to the herd instinct or the 'I'll have what she's having' *When Harry Met Sally* moment. But at the same time, we delight in the following exchange taken from *Monty Python's Life of Brian* – perhaps because we secretly believe we, too, could be Man in Crowd:

> **Brian:** You're *all* individuals!
> **The Crowd:** Yes! We're *all* individuals!
> **Brian:** You're all different!
> **The Crowd:** Yes, we *are* all different!
> **Man in Crowd:** *I'm* not...
> **The Crowd:** Sssh!

But taking risks is one of the things we should probably do a little more of. We don't take risks because we're afraid of the consequences of the risks – of the potential for failure, for disappointment. But in the failure and the disappointment there is also liberation and learning. Matthew Syed is the author of *You Are Awesome: Find Your Confidence and Dare to Be Brilliant at (Almost) Anything*, aimed at 9–13-year-olds. He wrote it to liberate young people from a fear of trying new things. Explaining his approach in *The Times*, he says, 'Lives are seldom destroyed by giving it a go and messing up; the silent killer of aspiration is the inability to even try.'[13] In his research, he discovered from teachers how 'the curse of perfectionism"* prevents risk-taking and learning. No one wants to look silly, a young teenager in a crowded classroom least of all. But making mistakes is the best way to learn, and holding fast to the status quo diminishes our opportunities to do so.

> In today's world of complexity and change, the importance of initiative has soared. The status quo is constantly being superseded. The defaults of today will be defunct tomorrow. This offers another reason why resilience is so important. For anyone who exercises initiative has to accept the possibility that things might go wrong. And yet these failures are not reasons to give up; they are opportunities to learn, and to pivot to a new way of thinking in a world that will never stop surprising us.[14]

Educationist Sir Ken Robinson sounded the alarm on this in a celebrated TED talk in 2006, when he said:

> Kids will take a chance. If they don't know, they'll have a go. They're not frightened of being wrong. And by the time they get to be adults most kids have lost that capacity. They have become frightened of being wrong. We run our companies like this. We stigmatise mistakes. And we're running national education systems where mistakes are the worst things you can make. And the result is, we are educating people out of their creative capacities.[15]

Shaking up the status quo and pushing ourselves outside of our comfort zones keeps us a little sharper. It helps us to pay more attention to those moments when we find ourselves following crowds in order to belong, or to make sure we don't stand out or

---

*   A term coined by researchers at Bath Spa University.

harm the group status quo. And this consciousness might give us the power to take an individual stand when we need to. It will also mean we can recognise when our instinctive status quo response might drive us to say yes (or no) when saying no (or yes) might be a more rewarding response.

Your status quo bias brain-rewiring plan

To make you more conscious of a natural inclination to keep the status quo – and to explore what happens when you occasionally break free of it.

Brain-rewiring reward

There is no doubt that the occasional shake-up brings some new energy and a little uncertainty into one's life. There is always an element of the unknown, a frisson of excitement, danger and adventure involved in countering this tendency to keep things the same, to keep the status quo – but that's what makes this insight fun. Understanding how this bias might be operating in your life and discovering how you can occasionally rock your boat will open new doors and create new opportunities.

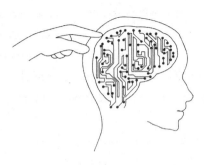

Brain nudges: Five Brain Team tips for engaging with the status quo bias

This section examines how the Brain Team started to recognise the ways that the status quo bias holds us back, enabling them to employ strategies that counteract it and reinvigorate their lives.

1. First, gain awareness or increased consciousness of how the status quote bias is operating and influencing your life

We all know it can be hard to throw caution to the wind; we worry that we might also be throwing the baby out with the bathwater. Status quo bias can do that to you. So, first recognise its influence; you can then decide what – if anything – you want to do with it.

Help! I don't like doing new things.

I realise I am not a risk-taker; that makes life stable, but less exciting.

I recognise in me a huge tendency to stay within my comfort zone. I like knowing what I'm going to be doing this time next week and I'm reluctant to be spontaneous.

I absolutely recognise that my default position is not to say yes and not to want to do new things. It's mostly due to inertia rather than fear of the unknown. Faced with an impending social event I always have a slight sensation of anxiety (even dismay) and a reflex preference for a quiet night in, even though I know I'm more than likely to have a good time.

It does make you think how differently things could turn out if you intentionally alter your attitude to risk and new experiences.

As consciousness builds, you can simply be open to being rocked – that is, you could decide in advance that you are going to be more spontaneous, should the opportunity arise ('Today I will say yes to whatever comes along…') – or you could proactively seek out the rocking and tell yourself, 'Today I am going to do something I have never done before…' Build space for a little spontaneity in your life and who knows where it will lead? Just being aware of places in your life (or in your head) that you might be able to shake up a little is a powerful conscious move.

Just thinking about rocking the status quo is a powerful behavioural nudge.

> *I love the concept. I hum it to myself and am just off out now to rock the boat – who knows what might happen? The world seems a big and positive place because of it.*

Don't worry: by challenging your status quo, you won't risk abandoning your core stability. *Occasionally* rocking your boat does not require you to shrug off *all* the aspects of comfort and stability you love in your life.

2. Gently rock the boat – a little rock can have a big impact

For the risk-averse, even little steps can add some element of excitement or difference. You don't need to take giant leaps – small ones will do the job. It's interesting to note that the most common actions our Brain Team decided on were those involving reconnecting with old friends. Even if the idea of getting in touch with an old friend pops into our minds and we think it would be a good thing to do, turning the thought into action is still hard; it is so much easier to procrastinate and put stuff off until another day. Many of the Brain Team bravely grasped the status quo nudge and used it to galvanise them into action.

I made a very overdue phone call to a friend in the UK and invited her to stay. She was over the moon and we had a great conversation reminiscing.

I rang two friends I had got out of touch with; I ordered a CD on Amazon by a singer I had never listened to before; and I arranged to have dinner with a new friend.

I called a friend in Newcastle who I haven't seen for eleven years, made spontaneous decisions to go for evening walks, visited friends and read my book in the pub.

Instead of taking the dog to the park, as I usually do, I decided to try a new walk along the river. It was utterly beautiful and I noticed the dog's tail was more waggy than usual, too!

I'd lost touch with a good friend who was my best man at my wedding; we'd slowly drifted apart. By ringing him and arranging an evening out, I created another positive, which resulted in a great evening.

I actually sent an email to two friends in the US whom I haven't seen for about nine years and haven't heard from for over two. I just thought, It's long enough, and I'd been putting off writing the message primarily because so much had happened and where to start! I can't tell you how good it felt to press the send button – it was like the proverbial weight off the shoulder; I hadn't realised how much the silence had been weighing on me. I find myself thinking of them often and am really looking forward to the reply.

I used business and social network sites and within a day made contact with a couple of old friends, one of whom I haven't seen for over twenty years. The good feelings from this made me determine not to let it slip again and also to make use of the sites more often.

3. Rock the boat a little harder!

Losing the fear of taking a chance is a great way to tackle the status quo bias head-on. When you're not afraid of the consequences and are prepared to let them unfold, surely something amazing is bound to happen...

I said yes to an invitation to a concert, although my instinct was to say no. The experience of listening to the string trio was sublime.

Being more conscious of our tendency to avoid things outside of our routine can make us more open to those very things. Think action with benefits.

It made me think twice before declining an invitation because 'That's when I usually do X.' will the world fall apart if I miss X this one week?

I was asked if I'd go to the reception for a project on Friday night, which was specifically to welcome the 100 kids involved and their parents. Well, that's exactly the kind of thing I avoid. I hate gatherings of people I don't know, and I would normally have made an excuse. But I said yes and forced myself to go up to strangers, introduce myself and get to know them. And I actually had a really good time – I not only made friends, but lasting contacts and very useful allies in the parents and teachers for the project. And I realised that most people in the room were even more scared than I was, and I could see they were actually very grateful that someone made the effort to speak to them.

After telling a friend about this project and this week's assignment, he asked me on a date and I've said yes, something I might not otherwise have done. So, who knows, maybe there will be an unlikely wedding as a result of this!

A rather wonderful example of breaking out of the status quo is to be found in the Art on the Underground Acts of Kindness project mentioned in the reciprocity bias section. In it, a man describes a status quo-defying leap-of-faith moment experienced years ago when, instead of conforming to the unwritten rules of London Tube travel – don't make eye contact, don't touch and for goodness' sake, don't engage in conversation – he rocked the boat:

> I saw the most beautiful girl on a crowded commuter train and she got off at the same station as me, Holborn. In the carriage I'd been reading a biography of Chekhov and had just read the words 'What you can do, or dream you can, begin it; boldness has genius, power, and magic in it.' I knew I'd forever regret it if I didn't speak to her and finally plucked up the courage on the escalator. Six months later we got married. We've now been married for fourteen years and have four children.

And that one rock might start a whole chain reaction and have you rocking and rolling through the day!

I did the cleaning and food shopping on Friday so that I could take my daughters to dance class on Saturday morning. In the afternoon, my stepdaughter called in unexpectedly with her baby, and instead of rushing round finishing off the Saturday household chores I would normally have started in the morning, I sat on the grass in the sun with the baby playing. Because of that, I didn't get to go running with my middle daughter, as planned, until 5 p.m.; and because of that, I didn't spend time making pizza dough as planned; and because of that, I nipped out and bought cheap pizzas for Saturday-night tea instead of making them; and because I went and bought pizzas, we ate late; and because of that, instead of eating round the table as usual (big family rule), we ate in front of So You Think You Can Dance *on the telly. What a treat!*

4. Try a status quo mantra for size

Use 'Rock the boat', 'Seize the day' or 'Take a chance.' Or you could try 'Do one new thing today.' Mantras are powerful, and they can help to empower us too.

At least two of the Brain Team used 'Rock the boat' as their personal mantra for the week; it helped one reach the end of a (hilly) sponsored walk, muttering *'Rock the boat'* through gritted teeth when the going got tough; the other brought the mantra into play each time she felt an impulse not to do something, so she would urge herself to *'Rock the boat – come on, rock it!'* and her week was busy and exciting. She was empowered to start conversations with strangers at gatherings where normally she wouldn't have, and took some significant steps towards a new career, which she had previously lacked the confidence to do.*

5. Stop your boat from rocking

For those who feel they already live their lives in turbulent waters, a little more calmness might be just the thing. You can also rock your boat by saying no.

Change is good in many ways but my life is so full of uncertainty and change that I would actually like more routine sometimes.

---

* Several months after the end of the project, the same team member told me she had started a full-time degree course and had set up and managed a unique retail concept at the same time as looking after her home and family. She said it was all down to 'Rock the boat'!

Reflections from the Brain Team: Increased awareness of the power of the status quo bias

There is no doubt that the occasional assault on the status quo can create new energy in one's life, and it has the potential to turn up the unexpected, to generate a little fun and a little uncertainty. It's definitely a way of widening one's horizons.

Do the unusual and unusual things will happen!

It helps to be a little more conscious than usual of one's comfort zones – the times, places and events where we subconsciously hold on to the status quo – and to understand how a single counterintuitive decision to make a simple change can influence everyone around us; and, like the ripples of reciprocity bias, how it can also set up a chain reaction of positive energy.

Above all 'rocking the boat' just made me stop and think before saying 'No'. It definitely made me pause and try to work out where my 'No' was coming from before actually saying it.

I can see that if we don't face up to change then life can be too routine and boring. The 'same old, same old' approach can lead to dull, dreary predictability. Small changes can give you the confidence to cope with bigger changes. Every journey is made up of just one step followed by another and then another and another.

The simple concept of maybe saying yes or no (as appropriate) when our usual modus operandi would have been the exact-opposite response undoubtedly generates energy and creates expectancy in the air. It is funny that just putting the suggestion out there is like giving permission. It can put a new spring in your step. Simply telling yourself you're open to new things and are prepared to ditch the default is enormously empowering. You get to feeling you could do almost anything! And remember, you are free to try new things and to make mistakes. Try again, fail better!

I have approached everything this week with much more gusto – slight risk, almost – and a sense of new adventure and new horizons.

Like many of the brain insights, you take control by countering some of the natural tendencies of the brain, so you might need to work hard to manage this and will probably need to keep giving your brain little jolts to jump start a new way of thinking. Find a way to nudge or steer yourself – a mantra or a Post-it note on the fridge can be very effective – to help build a habit.

To rock or not to rock, you are in control. It's your choice. The concept of rocking your boat (or someone else's) has an energy that (just like reciprocity bias) will cause ripples and knock-on effects. However, most of us need the comfort of the familiar around us, or the safety that comes from going with the flow of social norms. But just by being more aware of how the status quo bias operates in your life is empowering, and you don't need to rock. You can feel more aware and even desire certain changes and yet not act on them – that is also fine. We can make little tacking motions with our boat if we want, or rock it as hard as we can. It is up to us to decide how the insight will best fit into our life.

Brain-rewiring awareness: The Status Quo Bias

We do many of the thing that we do because experience has shown they work, we like them or they make life a little easier. We have created comfort zones in our lives and for the most part we are happy and relaxed within them. What this insight is really about is *understanding* that things that might challenge this state will be much easier to reject than accept, and that when we turn down an offer to do something that's new, it could be an unjustifiable response prompted by subconscious brain wiring rather than reflection. An increased awareness of this tendency means we can keep a door open to allow ourselves to embrace a little boat-rocking every now and then; to make more positive choices and not to put off or reject things simply because they don't fit into our established daily or weekly routine; to acknowledge that our lives don't have to be set in stone after all.

It's the little things again: from reconnecting with an old friend, to saying yes to something new once a week, to making little gaps for the non-routine – these small moments have the power to make life so much less predictable, so much more exciting. Also, in the wise words of journalist and blogger Esther Walker:

> To not change your mind about things, to simply blunder through life holding the same boring opinions and prejudices about the same boring shit, is awful and very ageing.[16]

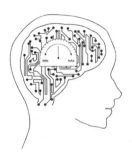

Summary of five brain nudges to engage with status quo bias

1. First gain awareness or increased consciousness of how the status quo bias is operating and influencing your life – this new awareness alone will make you think differently.
2. Rocking the boat gently can have big impacts, and we saw from just saying yes vs no, or picking up the phone and calling someone you lost touch with, can rock your world.
3. Rocking the boat a little harder could well be life-changing, as one of our Brain Team showed when increased spontaneity led to being asked out on a date!
4. Try a status quo mantra for size – use 'Rock the boat', 'Seize the day' or 'Take a chance.' Or you could try 'Don't be afraid to make a mistake' or 'Do one new thing today.' Mantras are powerful, and they can help to empower us too.
5. Stop your boat from rocking. For those who feel they already live their lives in turbulent waters, a little more calmness might be just the thing. You can say no to things.

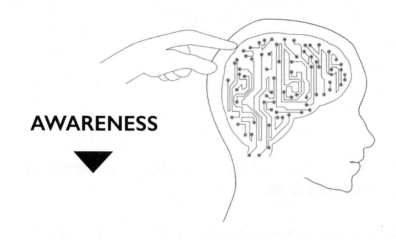

AWARENESS
▼

UNDERSTANDING
▼

ACTION
▼

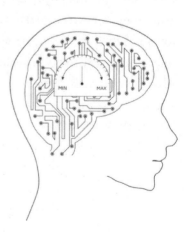

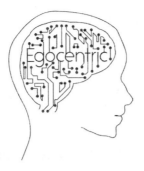

4. The Egocentric Bias

We're wired to be self-serving: How to make it less about you

Egocentric bias brain-wiring insight

We are all wired with a tendency to view ourselves and our actions through rose-tinted glasses, and this wiring is aptly named the egocentric bias. We gave our Brain Team the following insight to prompt them to consider the impact of the egocentric bias on their lives:

> We are all wired with a tendency to view ourselves through rose-tinted glasses. This means we might judge our behaviour to be more positive than it actually is. We are more receptive to and have a tendency actively to seek information that supports our own beliefs and to reject conflicting evidence. And, since it's very likely that everyone else will also be doing this, it's not surprising that there are clashes. We see the world from different viewpoints – mine versus theirs. We are also much more likely to believe that we are the ones who are right in argument/conflict situations.
>
> To add to the ME-centric bias we tend, when remembering past events, to puff up our roles in them: the fish we caught was bigger; we predicted a particular outcome when no one else saw it coming; we encouraged that person to behave in a particular

way and it saved the day; if only we had been in charge, things would have been done properly! This bias also plays into the rosy retrospection bias, the one that makes us view the past as a better place: the 'good old days' syndrome.

How the egocentric bias operates in our lives

Our brains are wired with a tendency to zero-in on our positive role in interactions with others. Egocentric bias also affects how we research things and, hand-in-hand with confirmation bias, it makes us more receptive to, and more likely to seek, information that supports our own beliefs, and to reject or simply overlook conflicting evidence.

The egocentric bias is what can cause us to regard ourselves as 'right', or a key contributor, in pretty much all the circumstances in which we find ourselves. It's more than likely that if you asked each member of a group (any group, it makes no difference) about their personal contribution to a decision made or a project run, the majority would imply that it was really their own efforts that made the most impact.

It's what makes us claim more responsibility for successes than failures and evaluate ambiguous information in a way that's beneficial to our interests. (Remember, it's also known as the self-serving bias.) Think about the following scenarios and you've got the picture. A student who gets a good grade on an exam might say, 'I deserved an A because I am intelligent and I worked really hard!' whereas the same student who does poorly on an exam might say, 'The teacher gave me an F because he hates me!'. Similarly, an athlete who does well attributes this to diligent training, fitness and having their head in the game, but blames bad luck, bad weather or a bad pitch if failure strikes.

In a study conducted in the late 1970s by Michael Ross and Fiore Sicoly, psychologists at the University of Waterloo in Ontario, results consistently demonstrated that 'one's own contributions to a joint product were more readily available, that is, more frequently and easily recalled' and 'Individuals accepted more responsibility for a group product than other participants attributed to them.'[17]

In their research, Ross and Sicoly explored the influence of the egocentric bias on the hierarchy of listing authors on the title page of a joint project – something they (as regular collaborators) had obviously had to deal with in their professional lives.

> One instance of a phenomenon examined in the present experiments is familiar to almost anyone who has conducted joint research.* Consider the following: you have worked on a research project with another person, and the question arises as to who should be 'first author' (i.e., who contributed more to the final product?). Often, it seems that both of you feel entirely justified in claiming that honour. Moreover, since you are convinced that your view of reality must be shared by your colleague (there being only one reality), you assume that the other person is attempting to take advantage of you.†

David Myers, a professor of psychology at Hope College, Michigan, describes a plethora of assertions prompted by the egocentric bias, and has this to say about its influence upon us:

> We more readily believe flattering than self-deflating descriptions of ourselves. We misremember our own past in self-enhancing ways. We guess that physically attractive people have personalities more like our own than do unattractive people. It's true that high self-esteem and positive thinking are

---

* For 'research' read endeavour, project or task, or putting-together of Ikea flat-pack furniture.
† The easiest solution, in an authorial context – should you be wondering what Ross and Sicoly decided upon – and one that requires authors to suck up their egocentricity, was to go with alphabetical order.

adaptive and desirable. But unless we close our eyes to a whole river of evidence, it also seems true that the most common error in people's self-images is not unrealistically low self-esteem, but rather a self-serving bias; not an inferiority complex, but a superiority complex. In any satisfactory theory or theology of self-esteem, these two truths must somehow coexist.[18]

The bottom line is we are hardwired with a tendency to give ourselves the benefit of the doubt, to take more credit for our successes and shoulder less of the burden of blame when things we are involved in failure. It's certainly something to bear in mind in conflict situations, because it's then that sticking firmly to the prompts of the bias will exacerbate things, while stepping back and taking a less ego-driven view might eradicate the flashpoints.

Although we should try to be aware of the bias in action and take steps to puncture it, there are times when a little bit of egocentricity can be a *good* thing. There is no doubt that the egocentric bias is part of our survival make-up – it boosts our self-esteem and evokes a sense of confidence in others who may be relying on us. And, you could argue, it is essential to our evolutionary survival blueprint. Think how much worse it would be if we were endowed with cognitive biases that ensured we always felt inadequate and always recalled our past actions as *less* successful than they actually were. We just need to know the egocentric bias is there and make sure we apply it or disengage it, as appropriate.

How the egocentric bias can work for you

Here's David Myers again:

> There is no doubt about it. High self-esteem pays dividends. Those with a positive self-image are happier, freer of ulcers and insomnia, less prone to drug and alcohol addictions. There is

also little doubt about the benefits of positive thinking. Those who believe they can control their own destiny, who have what researchers in more than 1,000 studies have called 'internal locus of control,' achieve more, make more money, are less vulnerable to being manipulated. Believe that things are beyond your control and they probably will be. Believe that you can do it, and maybe you will.[19]

Since the vast majority of us are more than a little in its thrall, we can leverage the egocentric bias in our dealings with others. For instance, if you needed to get some people together for a work, school or neighbourhood endeavour, you'd be more likely to recruit successfully if you approached people with egocentric-based phrases like: 'I was thinking, who is the most reliable and creative person I know?', 'I wanted to ask you first' or 'I was wondering who would be good at this, and you came instantly to mind.' Each of these speaks directly to a person's egocentric bias – whereas phrases like 'Loads of other people have already said they'll do it and I wondered if you'd be interested too?' (suggesting that the person you're talking to is way down the priority list) or 'I completely forgot about you... Don't suppose you'd like to help?' are unlikely to win any volunteers. And there are many management techniques that play to this bias – classic team-motivational tools that deliberately reinforce egocentric wiring by making an individual feel valued, important, even indispensable.

Just being conscious of 'me-centricity' can have a big impact on our behaviour – both the positives and the negatives. But for the moment, let's begin with these thoughts:

⊙ Imagine if we all listened a little more attentively to another person rather than trying to think of what we want to say next and when exactly we'll be able to shoehorn it into the conversation.*

---

* Lyndon B. Johnson, 36th President of the United States, got it dead right when he said, 'You aren't learning anything when you're talking.'

- ⊙ Imagine if we took ourselves off our pedestals even for a moment or two and allowed others to see us as we really are. It would mean admitting to weakness or failure, and it would mean taking genuine responsibility for all our actions.
- ⊙ Imagine if we didn't assume we were right and everyone else had got it wrong.[*]
- ⊙ Imagine if we could recall how much others have contributed to something, and how the success of a project isn't solely down to us.

Best of all, perhaps:

- ⊙ Imagine if we spent more time giving other people the benefit of the doubt, more time trying to put ourselves in their shoes so as to see things from their angle, and less time judging them from the perspective of our egocentric reality.
- ⊙ Imagine, imagine, imagine…

Your egocentric bias brain-rewiring plan

To increase awareness of the egocentric bias in operation in our lives and in those around us, and, occasionally, to try to short-circuit it.

Brain-rewiring reward

Just by being aware of the egocentric bias in operation in your life – in your communications, in your opinions, in your interactions – and trying to be especially conscious of it, you will become more sensitive to other people in many everyday situations. It will make you more open and draw other people to

---

[*] The dedication in James O'Brien's book *How to Be Right in a World Gone Wrong* (W. H. Allen, 2018), offers some helpful advice: 'For Lucy McDonald, who has taught me, among so many other things, that winning the argument doesn't necessarily mean you were right.'

you; it might make you more relaxed; it *should* make you a better listener and help to resolve conflict situations sooner.

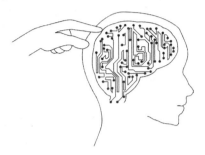

Brain nudges: Five Brain Team tips for puncturing the egocentric bias

This section shares the initial impact of living consciously with egocentric bias, the thoughts, feelings and subsequent behaviour it prompted, and a few strategies for keeping it in its place.

1. Remember, it's not all about you

However confident we feel that we don't have egocentric tendencies, there is always room for more deliberately conscious awareness – we shouldn't rest on our laurels. We need to keep this concept at the front of our minds on a daily basis. Acknowledging our own egocentricity, as well as that of others, and working around both, will make for less stressful encounters and will certainly prevent negative ripples from rolling out.

> *I've really noticed just how much I anticipate my next contribution – as well as noticing some huge egos around me who just can't wait to interrupt me!*

> *I have discovered this week that I am always interrupting people before they finish what they are saying. It's a terrible habit and I have vowed to try and stop it! It's partly that I feel I know what they are going to say next and am impatient to get on; and partly*

just that I'm desperate to interject my own thoughts. I thought I was quite a good listener – ha! In family life especially, it's so important to listen and not just go through the motions, especially with teenagers, who will sometimes drop pearls when you are least paying attention.

The deep observation of egocentricity in action both in and around you can be thought-provoking and a little terrifying. As you become attuned to it, you can start to see it everywhere. Even some of the life coaches in our Brain Team were taken aback:

It has made me realise how quick I am to bring myself into the centre of my thoughts, even though I'm 'listening'.

An understanding of how the bias might operate in others should help us to empathise better and be more able to take a moment to decode or reinterpret the influence of the bias in action.

2. Be a better listener

Before we can act empathically towards the egocentricity displayed by other people, we first have to identify its influence on ourselves. Thinking about our participation in conversation is a good route into this reflection. Is there *anyone* who can say they never think about an idea, anecdote or clever remark they're planning to add to a conversation, and are instead always listening 100 per cent to the person who is talking? A little self-examination on this tiny point can be enlightening.

The biggest insight for our Brain Team was realising how little people actually listen – and the Team tried hard to check this behaviour in themselves. Better listening and longer silences can lead to deeper, more rewarding and less stressful conversations.

I believe listening is a dying art, and in our present culture, verbal one-upmanship is all the rage.

It's a prick on my conscience about the times when I have been competitive or preoccupied with my own thoughts in conversations – anticipating my next contribution and not listening well.

Again and again in the past week, I felt the nudge on my shoulder of 'This is not all about you' and 'Just stop pretending to listen and really pay attention.'

The power of not listening enables one to misinterpret, get it wrong and miss the point.

It's hardest to dial down our egocentric bias with those closest to us because we tend to be in our most confident 'me' state then and it can be harder to reach solutions if conflict arises. It might well be that we are also at our laziest with those closest to us.

As a coach, I am trained to take the 'me' out of all client conversations and I like to think I achieve this most of the time when working. However, it was great to be reminded to think about it in my personal life and in my personal interactions. I had dinner with my husband this week and we got into a conversation about which of us is grumpier more of the time. It was very interesting to see how much the me gets in the way in these conversations, as, clearly, my hubby thought I was grumpier and more difficult than he was and I, of course, felt the opposite. I found by taking the me out of the conversation (or at least trying to), I was able to interrupt him less and feel more empathetic towards him, and it really made me try to see the situation from his perspective rather than my own (although I have to confess I still came out of it wishing I could help him take some of his me out of our conversations!).

I've tried to really hear my partner's argument, from an assumption that he might actually be right, rather than closing my ears to what he's saying, on the assumption I'm right.

It can be all too easy to multitask around family members and end up listening to them with half an ear. And sometimes it can be hard to make yourself heard:

My family are all shouters and interrupters, so it is jolly hard to get a word in edgeways. We are all on 'transmit' most of the time rather than 'receive'. It would be a revelation if we were all to reverse that, but I can't help feeling that if only one person (me!) did it, the others wouldn't even notice and would keep on shouting. But I have a feeling our family might be transformed if we could all achieve a little less egocentric bias!

Observing the bias in action can be fun and frustrating at the same time. Once you are more conscious of aspects of 'me, me, me' – such as the potential for competitive conversation – it can be quite fun to observe the bias in action in others, and it's probably a useful way of assessing your own egocentric checkpoints too.

I have a friend who's so egocentric it's not true! If I have a cold, he has flu; when I slipped a couple of discs last year, he went into enormous detail as to the number and complexity of the discs he'd slipped in the past. (You wouldn't believe the pain he suffered; it was so much worse than mine.) I read a book, he's read a better one – it goes on and on. I'm looking forward to seeing him again, now I know what drives him!

This insight is a salutary reminder that however brilliant we might think we are at listening and respecting other people's opinions or feelings, the truth of the matter is that we are constantly thinking of our own, and how to get listened to ourselves!

Playing to other people's egocentric wiring, by paying attention to what they are saying, can also be a powerful social tool and will hopefully stir up a little reciprocity bias in others.

The more you can really hear what the other person is saying, the more likely they will really listen to you. In the same way, the best way to receive empathy is to show empathy.

I definitely ended last week taking much more notice of other people and of how seeing things from a less 'me' perspective does make for a much more genuine flow of conversation.

A big takeaway is that to listen more and talk less, with longer silences, is a simple way to counter this bias. It's amazing how hard it is to listen without thinking of what you are going to say next, but it's worth the effort.

3. Come down off that pedestal – you're not so great

Another way of combating the egocentric bias, apart from listening well and better, is occasionally to present yourself in an unflattering light, because there's huge power in self-deprecation. Sometimes finding a way to proactively puncture egocentricity can be a useful tool. Telling stories where the joke's on you can work brilliantly in your favour if you want to put someone else at their ease or dispel awkwardness in an unfamiliar situation. Anchoring on stories that downplay or even undermine your role or contribution and offering others an unexpected reference point about you can shape the context for a more relaxed interaction. It's not easy to take yourself down a peg or two, but it's definitely better than someone else feeling the need to do it for you. (Word of caution: beware of deploying self-deprecation as an egocentric technique, i.e. *Look at me with my humility – what a big and noble person I am to share my flaws…*)

I began a difficult conversation with someone with a small self-deprecatory story about my trip in to work. I think it may have made me seem less authoritarian and scary, and helped us bond.

I have found that making jokes about one's own failings really does put others at ease. Amazing how it builds rapport quickly.

I chose this week to reveal the fact that my university degree is a 2:2! I have a reputation at work as an academic – I do have a PhD and I did go to Cambridge, but I only got a 2:2 when I was there. I chose to share this in a conversation and it was not easy as I am not proud of it at all, but I stated it and then resisted the urge to explain what else I had done at college to explain away the low degree. This was a humbling experience – and not one that I would readily repeat – but it did bring me down off a pedestal, so people are maybe now less threatened by me.

I have shared some of my less successful moments in life with the young people I work with, and this, I believe, has ironically made them respect me a little more.

4. Let go of egocentricity and give yourself a break

Life shouldn't be a competition, although your egocentric bias tells you that it is. Releasing yourself from the need to compete can be a soothing and enjoyable experience. It can be liberating to allow yourself to take a bit of a back seat and let others occasionally take on the heavy lifting.

I found it relaxing to sit back and let other people do the talking without feeling the need to interrupt and the need to concentrate on what I was going to say next. On several occasions I needed to bite my tongue when I was about to interrupt someone. By allowing the other person to talk more, our conversations had more depth and I learned things about various people which I had not realised before. Our conversations were definitely less superficial.

These last few days, I have learned a lot more about people simply by telling myself not always to interrupt their flow with an 'Oh, that happened to me, only worse/better...' story, but instead to keep listening and asking questions. This has helped me to remember more about what has been said, too, and I wonder whether actually listening properly to what people say will mean I'll now be able to remember their names next time I meet them!

This week I have really focused on listening to people and not switching off halfway through their conversation, and actually it is less stressful as your brain is only thinking of one thing, i.e. what they are saying.

5. Fight to keep your focus on dialling down the me, me, me

The underlying theme of this book is 'small changes can make a big difference', and you may need lots of small nudges to keep you focused on the goal of putting yourself second. Here are some of the Brain Team's ideas of small changes to use to support you in your deflection of the egocentric bias:

I'm trying to send emails that say, 'What do you think about...' rather than 'I think...'. It's a good start.

Listening and reflecting back responses, and questioning, is really helpful. If you ask people more about themselves, they are usually happy to expand. This can be really fascinating. It also puts everyone at their ease.

I prime myself before I start a conversation, saying, 'It's not about me.'

I plan to continue to try and see the world from another's perspective, and to say the words clearly to myself, 'He/she may have a point', more often.

I'll continue trying to put my ego aside and to remember that mine isn't the only – or even the best – perspective.

I will try counting to ten before saying what I think.

Reflections from the Brain Team: Taking 'me' out of the limelight

There is no doubt that simple awareness and self-checking can lead to major rewards in interpersonal relationships. Holding back a little allows you to be more present in the moment and more mindful. Among the benefits of confronting egocentricity are diminished stress points and conflicts, enhanced communications and improved relationships with family and friends. It's small steps and big wins.

Obviously if you can successfully do it you will massively reduce friction between you and others.

It'll make for more effective, and more enjoyable, relationships with colleagues and social contacts.

The process made Team members newly philosophical about life and what it is to be human. Living with the insight gave them time to think and reflect on this.

I think we are all incredibly self-absorbed and this is reflected in a multitude of ways. We get frustrated when we don't get our own way, when people jump the queue, and getting stuck behind slow drivers is everyone's nightmare. We all think our lives are pretty important and it is only when we let that go and let other people's lives affect us that we realise how much better we can be.

It has helped me grasp a real sense of the people around me, even those I thought I knew best.

'It teaches us how to be better listeners – to unlock and embrace the silence'

This is one of the big reflective thoughts to have emerged after a week's reflection on the influence of the egocentric bias. Brain Team members said that confronting it turned out to be a life-enhancing experience.

One of the Brain Team's revelations from attempting to listen better was the value of silence in a conversation; since people were not rushing to fill the spaces with talk, there were more gaps and breaks in their interactions. But it's not something that comes easily. A Chinese friend once said to me: 'The problem with the West is they talk too much and think too little.' We can be scared of silences, and silences in conversations are almost a taboo. Confronting egocentric bias seemed to address that taboo.

Create a mantra, a little conscious nudge – say less, listen more

In dealing with egocentric bias, reminders and mental nudges are essential aids for keeping us on track. Here are some suggestions for phrases you might repeat to yourself – all prompted by the experiences of the Team:

- *People love to be listened to.*
- *Just listen.*
- *I don't need to have the last word.*
- *It's not about me.*
- *It's all about you.*
- *I don't have to be right.*
- *I might have got it wrong.*
- *It was my fault.*
- *Silence is golden.*

Go on: write one down on a Post-it note – or invent one of your own – and stick it on the fridge.

Different times may call for different approaches

There is definitely a range of responses to the egocentric bias: some people will play it down, others will actively counter it and some may want to play the egocentric card more often. You need to understand when you might want either a little *more* me, me, me – to gain a little of the self-confidence it offers; or, if you're feeling a little under-appreciated or not listened to, when your mantra might well be 'This time it is *all* about me!'; or when you might need a little *less* egocentricity – to put yourself second rather than first. So be conscious of it in action but also fluid in your application of this brain-wiring insight; find your own goals and the tools to help you reach them.

One of the great discoveries for some of the Brain Team was that even though combating egocentric bias was hard to do initially, the more they focused on it, the more rewards it unlocked and the easier it was to put into practice. But not everyone achieves their goal – and when they're caught in the act, it's not a pretty sight:

What I have noticed more than usual – and it's something I tend to pick up on – is the worrying amount of people whose conversation is totally self-serving and who in virtually no way extend a thought towards their listener's world or sensitivities. We sat for three hours in the pub on Tuesday with friends we hadn't seen for thirty years. As we walked home, we discussed the incredible fact that neither of them had asked about us or what we are doing with our lives, beyond knowing we have four children. With egocentric bias firmly in mind, I was not prepared to talk about us anyway, but it might have been pleasant to arouse a bit of interest. What that evening crystallised was a) how hideously unattractive such ruthless self-centredness is, and b) that we'll make sure it's a good thirty years before we meet them again!

Brain-rewiring awareness: The Egocentric Bias

With our eyes wide open, we can see the egocentric bias in our own behaviour and in that of others. When we are more conscious of it in operation, we can be better receivers of information and also better at understanding a group dynamic.

The centre of much of what we do is conversation, and just having a little check on ourselves to listen more and interrupt less might lead to deeper and more rewarding communications.

Having an awareness of our tendency to rate our input as more valuable than an outside observer might means we should be more sensitive to or understanding of others. Another useful angle on the egocentric bias is the acknowledgement of a deadly default setting we might be prey to – the need to be right all the time, or perhaps more dangerously the belief that we *are* right all the time. We can trick ourselves into believing in our rightness because our perspective on the world is the one that matters most directly to us. But the more we try another angle on an idea or a problem, the more we step into other people's shoes, the more open we will be to new ways of thinking and doing.

A heightened consciousness of the egocentric bias (especially once we have taken the time to observe it in action, not just in ourselves but in others around us) can be extremely liberating.

Listen more deeply, be less invested in being right, and try to look at the world from another person's point of view.

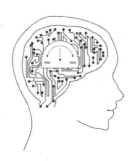

Summary of five brain nudges to deflate the egocentric bias

1. Anyone who feels they never or rarely display egocentric behaviour should, perhaps, ask themselves whether there is not something just a little self-serving about considering themselves completely untainted by egocentricity?
2. The biggest insight for our Brain Team was how little people listen – and they tried hard to correct this behaviour. A big takeaway is to listen harder and talk less, with longer silences. It can lead to deeper, more rewarding and less stressful conversations.
3. Come down off that pedestal – you're not so great.
4. Letting go of egocentricity can be restful. Releasing yourself from the need to compete can be a soothing and enjoyable experience.
5. You may need lots of little nudges to keep you focused on the goal of putting yourself second. The simple mantra '*It's not all about me*' might help.

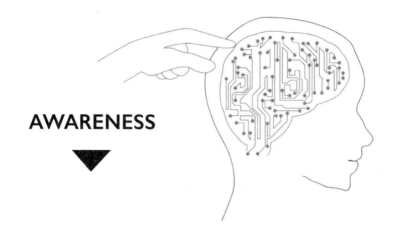

AWARENESS

▼

UNDERSTANDING

▼

ACTION

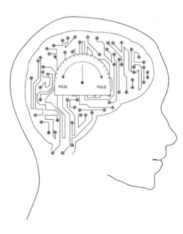

▼

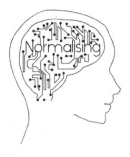

5. The Normalising Bias

We're wired to make the new old: How to rewind and refresh your life

Normalising bias brain-wiring insight

You can't help it: your brain is wired to make new stuff old hat – to normalise things.

> Habit converts luxurious enjoyment into dull and daily necessities.
>
> Aldous Huxley, *Point Counter Point*

We gave our Brain Team the following insight, encouraging them to see everything in a new light:

> We have an in-built tendency to normalise things. What is new and exciting one day can quickly become familiar. This means that the pleasure we get from things can be short-lived because we can't help but take them for granted. We are wired by the society we live in (and often by our peer group) to quest for the next exciting experience. Inevitably, the next thing also becomes old and lapses more into the subconscious, losing the excitement of the new. Once again we end up questing for more newness and the cycle begins again.

How the normalising bias operates in our lives

Our tendency to normalise things is a key part of evolution or learning and makes us more cognitively efficient, protecting our reserves of cognitive energy by programming us to pay more attention to what is new. So, like autopilot, it's a very useful aspect of our brain-wiring. But the problem is that it doesn't operate selectively. This normalising applies to everything in our lives – to people and to things. Everything is subjected to the same process, and our levels of enjoyment of some of the 'normalised' things in our lives can easily become diminished.

It's all down to our neurons. When our sensory cells are repeatedly exposed to the same thing, they stop responding to it – it's as if they get bored with it – which means *we* stop paying attention too. While countering our autopilot is all about building increased consciousness of the moment, tackling the normalising bias is much more about reawakening past feelings and fighting adaptive processes.

The first thing to be said is, don't beat yourself up about taking things for granted or forgetting how much you liked them; it's just how your brain is wired in order to make it as efficient as possible. In essence, we absorb, we adapt, and then, because of the way in which we live today, we seek out the *next* new thing – a kind of forward momentum that propels us in search of one new exciting experience after another.

You could say this is one of the curses of modern society – the one that tells us that unless we are reaching or grabbing the next thing and then the next, we are not growing or progressing, and there's a good deal of truth in this, especially in the context of ever-expanding new technologies. It encourages us to search endlessly for new entertainment – from leaping from one Twitter thread to the next or searching for a new Instagram discovery to add to our collection – creating in us an insatiable thirst for the new.

It is also a vicious circle because, inevitably, the next 'new' soon becomes the old new, and then just old, lapsing increasingly into our subconscious, losing the excitement it once had. And so again we end up questing for more and evermore new things, trapped on a ridiculous hamster wheel. At its worst it engenders a kind of permanent aura of dissatisfaction with our lot, prompting an illusory idea that new pastures will not only be greener but make us happier too.

The seductive properties of new are powerful, but new doesn't last long. Think about the last time you bought any of the following: a new car, a new dress, a new pair of shoes, a new blah... How long do you reckon you delighted in the wonderful newness before it became just a car, a dress, a pair of shoes? As long as we are seduced by new, we are prey to the vicious circle of dissatisfaction – our lives won't be complete without new stuff, so we get new stuff and then it just becomes stuff, so we need more new stuff and then that just becomes stuff... and so on and so on – for ever! Not surprisingly, the commercial world depends on the normalising bias and advertisements lure us to consume new products on a daily basis.

On her Twitter feed, novelist Marian Keyes taps into the joy of rejecting the power of new, having decided not to buy any new clothes for three months:

> It's funny, it's like a switch has been clicked and suddenly I'm horrified by the thought of buying expensive stuff, or new stuff. And it's been a lovely thing to fall in love again with my old favourites. I have lovely clothes and lovely things, and why not show them the love rather than chasing after the eternal new?[20]

Another term to describe the impact of the normalising process is the 'hedonic treadmill', coined by psychologists Philip Brickman and Donald Campbell in 1971.[21] They identified the disturbing concept that our brains cannot recognise an *absolute* level of satisfaction, and that no matter how much we have, we still want more. It's worth bearing in mind. Other academic works, including David Myers' *The American Paradox: Spiritual Hunger in an Age of Plenty*[22] and Robert Lane's *The Loss of Happiness in Market Democracies*,[23] have shown that happiness and fulfilment cannot be dependent on wealth alone. Sometimes, on hearing about the lives of the mega-rich and -successful, we ask ironically, 'Yes, but are they *happy*?' We know money can't buy happiness, but we find it hard to free ourselves from the belief that if we just had a little more of everything, our lives would be better.

In his 2019 *Times* article entitled 'If you think reality TV's bad, try the Game of Life', lifestyle journalist Giles Coren discourages the human quest for ever more wealth:

> You will never, ever think you have enough money. Because, the more you make, the more it will bring you into contact with people who have even more. Should you become one of those people who really do have more money than anyone else, you will discover that it doesn't make you happy and you'll wonder why you wasted your life in pursuit of it.[24]

In a research paper called 'The Psychology of Subjective Well-being', University of Illinois psychology professor Ed Diener and colleagues make the point that groups living a 'materially simple lifestyle', such as the Maasai in Kenya, the Amish in America and seal-hunters in Greenland, have positive levels of well-being, 'despite the absence of swimming pools, dishwashers and Harry Potter'.[25]

Whatever exciting and good things happen to us – a lottery win, a new car, a new job, a new house – these cause our happiness levels to peak. But these peaks will be only temporary, as we quickly become used to our new situation, taking it for granted and not appreciating it as much as we did when it was brand new.

Dan Ariely, psychology professor and author of *The Upside*

of Irrationality and *Predictably Irrational,* suggests that you can attempt to resist the hedonic-treadmill theory by investing in long-term memories rather than material possessions that you'll eventually cease to appreciate:

> If you are considering whether to invest in a transient (scuba diving) or a constant (new sofa) experience and you predict that the two will have a similar impact on your overall happiness, select the transient one. The long-term effect of the sofa on your happiness is probably going to be much lower than you expect, while the long-term enjoyment of and memories from the scuba diving will probably last much longer than you predict.[26]

Dan Gilbert, happiness researcher and professor of psychology at Harvard, has the same attitude:

> People think a car will last and that's why it will bring you happiness. But it doesn't. It gets old and decays. But experiences don't. You'll 'always have Paris' — and that's exactly what Bogart meant when he said it to Ingrid Bergman. But will you always have a washing machine? No.[27]

Psychology professor Thomas Gilovich and fellow researchers at Cornell University make the point even more forcefully in their paper on consumption and the pursuit of happiness, when they say we are what we do, not what we have:

> In a very real and meaningful sense, we are the sum total of our experiences. We are not the sum total of our possessions, however important they might be to us. If called upon to write our memoirs, it is our experiences we would write about, not our possessions.[28]

We are going to help you fight the normalising process in general, to be more conscious of adaptation and to anticipate it. Remember, though, that your brain's natural response will be to keep trying to normalise things, so you may need to develop a few little brain tricks of your own to maintain your 'renew and refresh' perspective. Think about behavioural checkpoints that will help you to avoid adaptation, or ways to keep your senses awake and conscious. Think about choosing *experiences* over things if you want to keep your hedonic treadmill covered in dust.

Your normalising bias brain-rewiring plan

To resist or at least check our tendency to normalise and adapt to people and things, and instead to seek to rewind and refresh the essential elements of our lives.

Brain-rewiring reward

The rewards that come from resisting adaptation and embracing the opportunity to rewind and refresh things that might have lost their sparkle are all bound up in keeping life and relationships new and exciting. And the first step lies in recognising how easy it is for things to lose their sparkle and then finding ways to fight this normative process. You can recognise the influence of this bias on relationships as well as on possessions and pay more attention to the people and things in your life you may have started to take for granted.

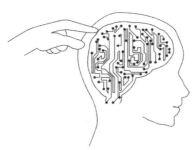

Brain nudges: Six Brain Team tips for countering the normalising bias and re-energising relationships with the familiar

This section looks at the impact of living consciously with normalising bias; the thoughts, feelings and subsequent behaviour it prompted; and a few strategies for keeping it in its place.

1. Make a conscious effort to derive new pleasure from what we've grown used to

When we open our eyes and look around at what we have, we can sometimes realise that we have forgotten to pay attention to it and therefore can't possibly be fully appreciating it. This isn't the end of the world, though, because once we become conscious of what we've been missing, we have the chance instantly to breathe new life into the things around us that matter and that mean something to us. This should help us to feel more alive and connected, and should bring a renewed sense of contentment and pleasure into our lives. It might also help us to think more consciously about how to keep aspects of our world alive a little longer by nudging us to hold on to the energy and vibrancy of all the new things; helping us to hold back adaptation for as long as we can.

> *I am noticing the good things in other people more (with some exceptions, of course!) and being more tolerant and noticing small things – the scent of lime, the smell of wet pavements, the poignancy and magic of dusk, the warmth of talking to a friend you trust, the beauty of my children still, when they are asleep; that wonderful first sip of cold sauvignon blanc as I scour the fridge for something that is not past its sell-by date to cook for supper.*

> *It has encouraged me to be aware of the fact that if something is amazing then it is always amazing – not just the first time you notice it.*

Not taking as many things for granted could make me more appreciative of my surroundings and make me embrace more of the things I love.

It has made me think about the first time I did things and the emotions I underwent then. Quite often, if something has had a strong impact on me, then that memory lies only just below the surface and is easy to revive.

2. We can let our relationships with people become monochrome. Think about how you can restore the colour

Taking the time to remember, reflect on and articulate what someone means to you is such a rewarding experience – for you and for them. We are sure that everyone is guilty to some degree of not spending enough time doing this. And the right moment to share your appreciation of someone else is almost always *right now*. Don't forget you can wait and wait for the 'right' moment and one day it might be too late.

This brain-wiring insight gives us permission to revisit all elements of our lives but especially the people in them, and it can help to bring new colour to relationships we have allowed to fade. You can make the familiar new again by asking questions and learning about unknown elements of people's lives. Ask close family and familiar old friends for stories about their youth, their first loves, their family feuds, their hopes and dreams, when have they been most thrilled, fearful, proud, etc. You get the picture. You'll discover amazing things, see them with brand-new eyes, and may even make new connections. You will also *most definitely* enhance your bond with them. A friend recently revealed that she rode the Grand National track aged seventeen and jumped Beecher's Brook in the process! It prompted all manner of questions and excitement and gave a new energy to our assessment of the friend.

Keep seeing people as the gift they are, especially the oldest and dearest friends that we presume will always be there. We don't know how much time we have to really appreciate somebody.

After thirty-five years of marriage, I am definitely guilty of taking my husband for granted and not noticing many of the thoughtful things that he does and the joy of having someone around to share the four Gs – good times, gut-wrenching worries, gossip and a good laugh.

With my parents in their mid-eighties, I realise that I am very lucky to still have them around, and that I can't take them for granted and should make more effort to spend time with them and show an interest in what is happening in their lives.

I also started saying a longer goodbye to my boyfriend, like we used to do before it became a normal thing. Just hanging round at the doorstep talking for longer really has reignited the feelings there.

It's made me take the time to look at everyone I'm in regular contact with (partner, daughters and staff) and try and see them afresh, reminding myself of their special qualities rather than their shortcomings.

We may well need a little behavioural nudge to realise what is important around us – while we still have time to do something about it. Taking for granted the people and things in our lives is all too easy in modern, 24/7 life. This was particularly poignant for two Brain Team members. One had a friend who had recently died aged only fifty, and the other a friend who was terminally ill. The resonance of appreciating what you have when you have it sang out in their reflections:

> *Strangely enough, this brain insight has been so relevant to the sadness we have just endured. Losing a very dear and close friend at fifty years to brain cancer has been an experience not to be repeated. However, the insight to rewind and refresh has been an intrinsic part of our grieving and recovery. It has made us all realise that we must kiss our partners just a little longer, we must dwell over those photos and share them and ring rather than text or email. It is amazing how we take our friends for granted – they are here but then they are gone.*

> *I was with a friend today who has pancreatic and bone cancer. She has had a stomach bypass operation and is now having chemo in order to prolong life, knowing that the cancer is terminal. It was the first time I had seen her since the diagnosis. She didn't want to talk about her condition much but she did say that she is taking each day as it comes and relearning the habits of a lifetime by never planning more than a week in advance. She said that she had agreed with her husband that each day she will do what she wants to do. If that means sitting in the garden and just enjoying it after years of tending it, that is what she'll do. She said she didn't want a holiday in Bali or anything dramatic; she just wants to really enjoy and take notice of the things that really matter to her. To me these seem to be the simple things in life – her children, husband, home and close friends. She said she wanted a sense of normality, for the children's sake as well as hers. I thought of this week's 'brain insight' as I was reflecting on our conversation. Things are horrendously difficult for her and her family. What I see is that she is finding a kind of peace in her focus on the here and now, on what she has, and in experiencing afresh each thing that she already has.*

We also love the idea of not only thinking about what you might feel about someone, but also thinking about how you might *tell* them what you feel.

> *I thought about each person's skills and then thought about if a little nice note is left for them every now and again, how much better that would be.*

3. Reawaken old memories – think about the stories contained in the things around you. It's a powerful way of renewing deep thoughts and feelings

Revisiting old memories can prompt increased feelings of contentment by reminding us of the life story we have built around us, and bringing to the surface the richness of all that we have, all that we have had – the people in our lives, past and present, the places we've been and the things we've done – is something to savour.

> *It has proved a lovely time to reminisce. We have a fabulous large original 1972 painting of two young girls running into the blue, blue surf. I remembered first seeing it when we entered the shop. Oh, it was love at first sight, as it took me straight back to being eleven years old. I would go as far as to say that I thought I even believed it was a portrait of my twin sister and me at the beach. We spent most of our summer hols at nearby beaches in Wales. It made me think of hot days and the feel of our funny bobbly, stretchy costumes, ham sandwiches and the smell of calamine lotion. Delicious heaven.*

> *I became much more appreciative of my life and my things. I polished a much-loved table – usually unheard of! I looked at all our photos in frames and rearranged them, but not before I had relished the old photos of happy times.*

There are pictures that have hung on my wall for years and little figures I have bought on holidays that, when I got them, meant so much to me, and they had just become background objects. But I spent an afternoon this week cleaning them and really looking at them and the memories flooded back; some made me laugh and some made me cry, but all for good reasons.

We created 'the album of the week' and put it on the kitchen table. What is amazing is how the teenagers loved flicking through and remembering times past and old relationships. Sometimes it is also a reminder of closeness and fun times, which as we know can easily get forgotten. So simple, but so effective.

I revisited one of my digital photo files and relived some very happy memories. There, smiling back at me and with arms entwined about each other, were so many great friends at a birthday celebration. Great photos all round, and I started to recall things with pleasure, like the pot I painted for a friend that week for her birthday present.

Normalisation happens to all of us, and it's unavoidable. If we lived our lives as if we were always seeing and experiencing everything for the first time, it would be unbearable. However, there is definitely a value in polishing up some of the original buzz from time to time. Try to remember the excitement of a first kiss with your partner, the moment your baby was born, the day you moved into your new house, the thrill of hearing you'd got the job, and you'll get the idea. Think about the things and people you may take a little for granted. Look around your home at pictures and photographs, at favourite items of clothing. There's so much tied up there.

4. Awaken newfound appreciation by depriving yourself of something you take for granted

Think how good that coffee will taste after a couple of days of abstinence. Give up wine for a month and imagine how you'll savour that glass of perfectly chilled sauvignon blanc when you eventually choose to have it.

5. Invest in experiences that have a rich and long-lasting value, and that can set up a whole pattern of sustaining memories

Think about *doing* stuff rather than getting stuff.

6. Make a list of the things that you take for granted but would like to refresh

> *I tried to make a list of things that had lost their sparkle and what I could do to revitalise them, e.g. reading the papers at the weekend for the pleasure, not just for info or for a sense that I have to read them as they've been delivered; enjoying the garden for its beauty and smells rather than the need to weed the brassicas and pick the last of the strawberries.*

Reflections from the Brain Team: Challenge the normalising bias for a gift that keeps on giving

The real strength of this brain-wiring insight lies in its simplicity; we all recognise the thought and understand it, but it's often hard to act upon. Sometimes we just need a nudge in the right direction, and then the rewards for even taking a small step are mighty.

One Team member reframed the fight against the brain's adaptive/normalising nature as 'the gift of giving oneself the space and time to remember'. It's a good thought – a mantra to work with.

> *More and more I realise in life that it is one's attitude that is important; that we have a choice and we can see things, at a basic level, either in a positive, or a negative, light – typically, glass half full or half empty. Self-reflection and a checking process – which these insights encourage – help us to slow down, to look around and realise what we have and where we are, and to find joy and contentment. We just need to take the time to look.*

Refreshing our view of those close to us

Like reciprocity bias, this is hugely enjoyable to put into practice. Awakening our appreciation of how much we value the people around us is one of the simplest and most extraordinarily rewarding things this brain-wiring insight can help us to achieve. Get your inspiration from a twist on the Capra movie *It's a Wonderful Life* and ask yourself, What would be different about my life if the people around me weren't in it? Planning future actions to appreciate and value these important people will make you feel good too.

Old really can be the new 'new'

Wearing a metaphorical anti-adaptive hat, it's easy to relish everything anew: pictures on a wall, old photographs, favourite pieces of clothing not worn for ages, buildings routinely passed, the aroma of a freshly brewed cup of coffee or a newly mown lawn. Small steps – big revelations. In general, this insight brings with it something like a sense of relief – of gratitude, even. For the Brain Team, it complemented the idea of appreciating and really living in the moment that had been prompted by the great autopilot switch-off. It offers permission to be content, and to enjoy, without yearning for more. And it's bang on trend for a world drowning in overconsumption.

> *This week's task has really had a positive impact on me; I now go into my house and smile when I see certain things just bringing back memories. I'm aware too that this insight could have helped me in the past to remember things and not just discard them, reminding me of what certain things and people mean to me.*

> *I have really enjoyed this week's insight – looking at things and people in a 'new/old' light and remembering why I have them and what they mean to me. The simplest thing I have done this week is spending time with my mum. Although I see her all the time, I take that for granted, so this week I really took time out to listen to her and do things with her which brought back how much she means to me. Now she isn't 'just there' any more but she is there because I want her to be.*

There is no doubt that rewinding, reminding, renewing and refreshing one's life architecture seems to deliver a level of deeper contentment for many. And it doesn't take much – even five minutes each day is enough to pause and refresh.

And overall, it really is a wonderful life

The beauty of retro steps is all the associated feelings and memories they can evoke. This is the magic of the brain and how it is wired. By exciting one area, you will connect to others, and so on.

Brain-rewiring awareness: The Normalising Bias

In our 24/7 society, which encourages us to focus more on the future than on the past and present, here's a quick route to heightened awareness:

⊙ to open our eyes and look around at what we have and might be failing to appreciate

⊙ to take the time to remember why we love the people and things in our lives and to be more active in our enjoyment of them right here, right now

⊙ to consciously fight the continual brain circuitry working to normalise and adapt to the new; to fight adaptation so as to hang on to the zest and energy of things, people and places (old and new) for a little longer, to enjoy their magic and spark

> *I feel I've got 'permission' to think, to refresh, to reflect, and I feel so much better for it. It's dreadful how easily we can allow ourselves to forget what we have, to not give ourselves the time to look at what we have, to value what we have forgotten.*

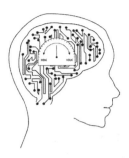

Summary of six brain nudges to combat normalising bias

1. With conscious effort, you can derive new pleasure from what you've grown used to.
2. It's easy to let relationships with others become monochrome. Think about how you can restore the colour.
3. Reawaken old memories – think about the stories contained in the things around you and create a space for yourself to look, think and reminisce.
4. Awaken newfound appreciation by depriving yourself of something you take for granted.
5. Invest in experiences that have a rich and long-lasting value, and that can set up a whole pattern of sustaining memories.
6. It will help if you make a list of the things that you take for granted but would like to refresh; the process itself will encourage you to take action.

AWARENESS

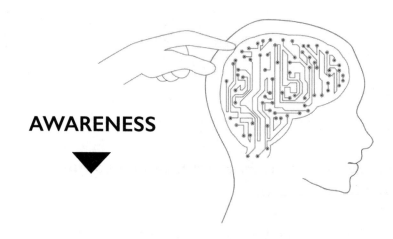

UNDERSTANDING

ACTION

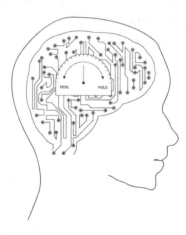

6. Framing and Chunking

Mental strategies that impact on our ability to receive, interpret and manage information

Framing and chunking brain-wiring insight

Behavioural science has shown how our brains are incredibly sensitive to the ways in which information is presented to us. We gave our Brain Team the following insight to encourage them to interpret information and events differently by applying new ways of framing:

> Framing is a powerful tool – it can change behaviour and make a difficult task more achievable, and it can also impact the way information is processed and understood.
>
> If you actively try to think differently about something you can bring about a change in the whole experience. This is the beauty of how our brains are wired – we have more control over how we conceptualise an experience than we realise.

How framing and chunking operate in our lives

You can choose to look at them in two ways – as a strategy to trick our brains into seeing something in a different light, *or* as evidence that our brains are clever enough to adapt the idea of a task – the thing you have to do; the information you have to interpret – in order to help us see it as more achievable or to achieve it. Whichever way you choose, it works.

Framing – a brief reflection on what it is and how it can operate in your life

All manner of experiments attest to the significance of framing in its various forms, whether that involves presenting or interpreting information differently, or priming people to respond in a particular way. In other words, if you want someone to believe something or behave in a particular way, science has shown that you can do it (and you can use it on yourself, too).

Here's a simple example. Imagine you were about to have an operation and the surgeon told you, 'Out of every hundred patients having this operation, ten will be dead in five years' time.' It would sound pretty alarming. If, however, she presented the information like this: 'Of one hundred patients having this operation, ninety will be alive in five years' time', it presents a much more comforting prospect. If you were the patient who learned about the ninety survivors rather than the ten unfortunate fatalities, you'd probably be more likely to agree to the procedure. We know that we tend to overfocus on the negative – it's the negativity bias at work – so how you present the information is critical.

Here's another simple but incredibly effective demonstration: 'Twenty-five per cent of people don't like X' can easily be flipped to 'Seventy-five per cent got real value out of X.' Easy to see which is more appealing. It works on an everyday level, too. If you present consumers with meat labelled '75% lean', most will assume it will be tastier and better quality than the identical meat labelled '25% fat'.

Once you start to open your eyes as to how you might reframe or reconceptualise the things you have to do or ask others to do, framing can take on a life of its own – it can even become addictive. Start with things you do on a daily basis. None of us really likes commuting for its own sake, but why not try to reframe, or rethink, your commute on the train or in the car? Think of it not as a waste of time, but as your *private* time. What could you do to get more out of this time? What could you do that would ensure you actively looked forward to this time – made it *your* time? Imagine you started reading or listening to your favourite book or music, *but only when you were on the train*. You might find yourself longing for that trip! Or when you drag yourself out of bed and think '*I must go for a run*' or '*I must go to the gym*', maybe it would help if you were to start thinking about a session in the gym or an early-morning jog as '*a gift to my body*' or '*my thinking time*'. These may sound like simple ideas but building a different context around things – a positive aura, if you like – can be incredibly powerful, and our brains react wonderfully well to this.

Whenever we interact fleetingly with strangers, we automatically apply a frame of some kind to them. Often it's a negative frame (remember the egocentric bias with its 'we're right and other people are wrong' compulsion?). Think about it: the guy who overtook you on the inside lane of the motorway, the woman who barged you with her shopping trolley and didn't apologise, the group of teenagers who walked past you in the street talking too loud and smoking – we frame them: the guy was an idiot, a megalomaniac; the woman was arrogant, snooty; the teenagers – criminals in the making. But maybe if we took a moment to *rethink* the framing we apply so instinctively (and egocentrically, too), we might find a more compassionate angle: the speeding driver – rushing home to a sick child? The 'arrogant' woman – recently bereaved? The teenagers – anxious about exams and job prospects? If you think about it, letting them off the hook liberates you too, because it frees you from anger and irritation, and engenders a more sympathetic response.

Look at these examples to see how framing can be applied, whether it's evaluating our own lives from another person's perspective, judging our successes in part on lack of failure, or contrasting one uncomfortable reality with an even less appealing alternative:

⊙ A woman in her late forties was talking to her oldest friend. She drew comparisons between their lives. 'Look at you: you've got an amazing job – you write for a great newspaper – as well as having two young children, while I'm still struggling to get some kind of ambition in my life. And my kids are now teenagers who don't want to spend any time with me.' The friend looked at her in astonishment. 'You're kidding me? I've always thought your life was so good. You have such freedom and such great friends. Your children are really independent and in control of their lives. What are you complaining about?'

⊙ A girl and her sisters would regularly reproach their mother for what they regarded as her failure to appreciate their worth (for instance, when she expressed dissatisfaction about a 65-percent exam result they had got, or a netball match they had lost). 'What are you complaining about?' they would remonstrate. 'We're not drug addicts! We're not pregnant at fourteen! Why aren't you more pleased with what we do? Give us a break!'

⊙ A mother of a very sociable teenage daughter, who was forever out at parties and sleepovers with friends, asked herself: 'Would I prefer it if she were home with me all the time, with no one calling to ask her over?' It allowed her to feel better about her daughter's busy lifestyle.

Chunking – a brief reflection on what it is and how it can operate in your life

You know chunking when you see it. It's how the media regularly serves up our news and entertainment. Every morning the BBC News app on my phone leads with 'In a hurry? Here's what you need to know in five minutes', and you'll most likely have had your fill of articles on the 'Top 5 Places to Find Winter Sun', '10 Best Spa Breaks' and '5 Ways to Reduce Belly Fat' (that last one especially). It's all chunked for our easy consumption, and it works. We are drawn to the ease of such listings. It's good to know, then, that we can make that same sense of ease work for us in other, perhaps more relevant, contexts. That Top 10 list mentality can work brilliantly if you are trying to explain something or argue a point – for example, 'If you want to understand X, you need to know these five things' or 'There are three key things I want to say.' It's a powerful approach. Listen to politicians with their dominant 'Firstly...' as they take charge of debate.

We can easily be daunted by big tasks. We can overfocus on the end goal and see it as so huge and unattainable that we might as well just give up before we even start. But our brains are designed to help us to reach our goals, as long as we point them in the right direction, and one of the tools we have at our disposal in working towards our goals is chunking – a key element of framing.

Chunking encourages you to break things down into chunks, or separately achievable parts of a whole. It means that instead of thinking about the end goal, you focus only on the first step and then the step after that and the step after that, and so on. It's incredibly simple but it plays to the brain's strengths in helping us achieve amazing things. In June 2019, Selah Schneiter, aged ten, became the youngest person* to climb El Capitan, the 3,000-foot rock wall in California's Yosemite National Park. Recalling the climb, made with her dad and a friend, she said, 'Our big motto was, "How do you eat an elephant?" Small bites.'

We have the power to interpret and perceive the same information differently depending on how we frame – envisage or picture

---

* A few months later, Schneiter's extraordinary achievement was superseded by that of Pearl Johnson, aged nine!

– it. As a result, *framing* is an incredibly powerful tool. Change the frame and you can change the meaning and the behavioural outcome. It can be the difference between taking one course of action or another. Meanwhile, mental 'chunking' – breaking down previously daunting goals into manageable steps or chunks – can make any kind of behavioural change much easier.

These two concepts are different from the various biases in that they focus on how the brain reacts differently depending on how the same information is presented. Just a slight change in presentation can create a completely new perspective, a different interpretation and an alternative outcome. Whether simply making something seem more fun or breaking a daunting task into achievable steps – chunking it into parts – the mental frame changes the picture.

The brilliant thing about both framing and chunking is that they are easy to do in tiny steps, and we all know by now that tiny steps can make a big difference.

Your framing and chunking brain-rewiring plan

To use framing and chunking to effectively trick our brains or encourage them to engage with things from a different perspective.

> What a wonderful life I've had! I only wish I'd realised it sooner.
> Colette, French author (1873–1954)

Brain-rewiring reward

Being more conscious of the power of these brain strategies will make you more sensitive to how you interpret or present information or how you approach a challenge. They will give you the power to make a bad thing better, a hard thing more achievable, or what can be seen as dead or wasted time as time better spent – that's framing and chunking for you. They're useful tools to have at your disposal.

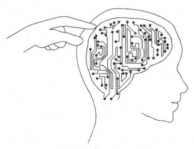

Brain nudges: Six Brain Team tips for getting the most out of framing and chunking

This section shares the initial impact on the Brain Team of living consciously with framing and chunking, the thoughts, feelings and subsequent behaviour they prompted, and a few strategies for applying them.

1. Remember that small changes in how information is presented or received can create a completely new perspective

Having mental framing at our disposal will make us better coders and decoders of information.

> *So often a situation can appear dreadful – a worst-case scenario of imagined possibilities, with fear and uncertainty swiftly stepping in. Yet, with just a little rethinking, there is often another perspective – another scenario, alternative thoughts to be entertained – that helps to reduce or remove the fear, facilitates coping, moves us forwards towards a solution, resolution or acceptance. Paradoxically, there really is so very often a silver lining to every cloud.*

There's an enormous appeal in the idea of taking something not so good and making it better. It offers a sense of liberation and promise. Where can you apply it?

I tried to get into the habit of automatic reframing when I thought something negative about a job or a person. Instead of being just a taxi service to the children, it gives me the opportunity to talk to them on their own, and share difficult things sometimes, because the car is somehow less threatening than face-to-face.

I'm suffering horrible pain from a frozen shoulder; it means I have to go to London for cortisone injections, which is a bit of a schlep. But it means I'm able to schedule a lunch with my daughter, so I'm focusing on that side of things. Less pain, more lunch with my lovely daughter!

2. Choose a frame that works for you

What is particularly rewarding about this brain insight is the amazing diversity of ways in which you can use it. The process not only allows you to achieve tasks but also to gain greater control of the narrative.

I have decided to start to use the rowing machine we have and to listen to an audio book only when I row, which is great! I've reframed it to row while I listen instead of the other way around. And it makes me stop watching the timer, which I would otherwise have my eyes glued to!

I have seen my journey to work as quality time to properly talk to my mum (who I travel with most days), rather than just a drive. Really listening to her and taking it in, not just hearing her.

The precious 'reframed' bit of my day is my Tube commute — my husband bought me a Kindle and, ever since, I've eschewed newspapers and instead I'm determined to read every day. It's half an hour there, half an hour back, so that's ten hours' reading a week, and a very good use of otherwise dead time.

3. Use reframing to put yourself in someone else's shoes and give them the benefit of the doubt

Next time someone is rude to you, do a simple reframe by asking yourself the question, 'What awful thing has happened to them to put them in this mood?' (Not only will you relax a little, but it's also a great way to avoid being overwhelmed by a negative wave.) I used just this kind of reframing during one of those irritating call-centre experiences the other day. I had phoned to make an automatic payment and wasn't expecting to speak to anyone, but for some reason I was put through to customer services. The man who took the call sounded incredibly unfriendly; I felt my hackles rise. I was almost ready to respond in kind (reciprocity bias in action) when I decided to reframe him. It imagined it was his first day. He was reading from a script (hence the unfriendly nuances I was picking up) and his supervisor was breathing down his neck. He was having a tough time of it. I changed my tone, I smiled when I spoke and even attempted a bit of a joke. He relaxed; his voice altered straight away. He laughed at my joke. The call ended well.

When people say something to me and I feel a negative response inside, I do maybe pause now and say, OK, well, I could reframe that as X, and this often takes the tension out of my first reaction.

I've realised lately that when I think other people are unfriendly, they might well be responding to me! I'm not the cheeriest of souls.

4. Conceptualise reframing as a simple piece of life-armoury you can carry with you and pull out any time you need it

The ability to reframe on demand is a powerful coping skill.

An area where I have been more successful is to use the time when clients are late for an appointment to just quietly catch up on my list of things to do, reply to a few texts and generally sit back and give myself a moment to breathe – so much better than sitting and fuming as to why my client is late! I am very keen to keep this going and to really keep up the reframing – sounds like a much less stressful way to operate!

I have increasingly found that it is our approach to things in life that matters more than what actually happens, or rather that we can change our experiences of things by looking at them differently, by framing them in different ways.

After reading [Oliver James' book] Contented Dementia, *I was encouraged to view each encounter with my mother as a game, so that I would go along with her conversations and humour her rather than questioning or correcting her. Doing this meant that the time I spent with her was less daunting, although still challenging – but in a more positive way as I would concentrate on talking about her earlier years and her hobbies.*

5. Use reframing to engender more positive interpretations of the past

It will definitely help me to see change, which I might once have seen as 'loss' or 'failure', as growth, learning experience, opportunity to share things with others on the same journey. It will thus, I hope, enable me to welcome changes and events, and not see them with trepidation.

If I had had this perspective when my last relationship broke up, I'd have not dealt with it as 'rejection' – I'd have realised it was my then partner politely sodding off out of the way to make space for the lovely chap I now have in my life! When something is lost, it truly is making room for something better to come into your life.

6. Take it step by step and break it down. 'Chunk it' is a great mantra

Here's a simple example of how to chunk something, which literally takes one step at a time. Say you have to carry an incredibly heavy suitcase up a steep flight of stairs. You are standing at the bottom of the stairs and you are looking all the way up to the top. You are daunted by the task ahead, thinking you will *never* manage to carry the case all the way up there. Now reframe the task and *chunk it*. Look only at the first step and tell yourself the task is to get the heavy suitcase up just one step. Then you look to the next step and your task is instantly achievable: one step at a time.

Here's how you can use chunking when taking exercise. This example was provided by one of our Brain Team, who swims regularly but sometimes reluctantly. Her end goal is usually seventy lengths of a twenty-five-metre pool. Once she's in the pool, she thinks in chunks of ten lengths at a time: 'Do ten lengths and then twenty is in reach; do twenty and thirty is achievable; after thirty it's a snip to get to forty; at forty I might as well do fifty; at fifty, why stop now? And at sixty, I'm just ten laps away from a shower and a feeling of smug satisfaction.' She can alternate the length-chunking to think in terms of time, too – say it takes her five minutes to swim ten lengths of the pool and she gets to sixty lengths and desperately wants to finish, she might tell herself, *'It'll only take five minutes. What am I making such a fuss about?'*

For the chunking to work without making her feel like she's on a treadmill she can't get off, she also gives herself 'get out of jail free' chunk cards; so, she might say, *'Well, if I get to thirty and the pool's too busy, then I can leave'* or *'I don't have to do seventy today;*

I could just do fifty and then seventy tomorrow.' Simply having the imaginary permission slips allows her to relax; there's no panic, it's not urgent, she can stop if she wants to. More often than not this means she swims all seventy lengths without taking a break.

My mother used a wheelchair in the last years of her life and was on oxygen twenty-four hours a day. She lived in a sixteenth-century cottage, up a steep drive, with two flights of stone steps to get to the front door. She had two automatic stairlifts to carry her up the steps but had to walk from the car to the first stairlift and from the second stairlift to the house before she could sit down and recover. She couldn't help but be completely daunted by the prospect of making this journey; it was almost enough of a threat for her to consider never leaving the house again. Then I suggested 'chunking' it. I put extra chairs along the way and encouraged her to break the journey down so that she only ever had to envisage making the journey from one chair to the next. And it worked. One chair at a time.

Think of the stuff that you know you have to do but wish there was a way of enjoying it more or getting through it without having to grit your teeth. Begin to think if there is any way you can reframe, re-present, chunk and revalue those tasks.

The thing with chunking is it's already part of your mental programme. Think about it: there's 'one step at a time', 'one day at a time' – it's all chunking and it's already hardwired into us. But we don't always use it as proactively as we could. However, like the Brain Team, you might find that as soon as you've got mental chunking in your head, you'll suddenly find yourself saying 'Let's chunk it' more often than you ever thought you would. The Team certainly found chunking one of the easiest techniques to apply and they used it in all sorts of contexts: for health and housework, for exercise and for work-related tasks. They cleaned their house thinking *'One room at a time'*; they decluttered stuff into the attic thinking *'One box up equals one reward of a look through old stuff while I'm up there'*; they got through difficult hospital treatment sessions by visualising them in thirty-minute stages – *'Much easier than thinking "I'm going to be in here all day and I can't believe we haven't even got a cannula in yet and we've got to wait for a doctor to come etc., etc.,"'*; they got their children to finish their breakfast *'one bite at a time'*.

Reflections from the Brain Team: You can reframe anything

Find ways to practise these winning brain strategies and keep them to hand, ready to deploy when you need them. Create your own internal mantras: *'Let's reframe that'*, *'Look at it another way'*. However you choose to do it, it will certainly pay dividends, and the more you use it, the better you'll get at using it most effectively. Even thinking about reframing a situation or chunking a task will, at the very least, give you pause, time to take a breath and reconsider.

> *This insight had the most dramatic effect. I found that it was easy to embrace all of the things I sometimes frame as extra pressure, like going to the gym or planning the weekly meals, and transform them into thoroughly worthwhile (even enjoyable) activities.*

Fun is a particularly useful reframing tool

The simplest type of reframing is to take something that's dull in some way and make it fun. There have been lots of ideas over the last year or so to increase happiness (see, for example, author

Gretchen Rubin's brainchild www.the-happiness-project.com) and an injection of fun has often been central to these. It is just one of many ways to change the frame but it's worth an exploration on its own.

To most of us, the most familiar example of fun framing in action is probably a child's plate of healthy food (almost certainly including broccoli) with a child who's reluctant to eat it. The fun element, as we all know, involves reworking the dull food components to represent a smiley face – with lots of bushy green hair – or a forest of little trees, all the better to encourage the child to tuck in. But that's just the beginning. Why should the fun framing end there? Experiments have shown that fun framing can be used to make people act in a socially responsible way without their ever feeling disgruntled about it, and, in fact, rather enjoying themselves. Speed-limit signals that smile at us when we are within the limit; litter bins that have been adapted to make a falling and splashing noise to reward us for picking up and disposing of litter; stairways remodelled to form giant piano keys that 'play' as we step on them, to encourage us to take the stairs rather than use the escalators – all of these are brilliant examples of the way fun can guide us to make the right choice. There's even the road design that plays music as a car's tyres make contact with the specially treated road surface. However, if you drive too fast the music's tempo goes haywire; it only plays to the right time when you stay within the speed limit.

Think about where you can see fun framing in action around you and how it is being used on you. We are completely familiar with fun in the context of *fun*draising: Comic Relief applies it wholesale; fundraising fun-runs where all the competitors wear Santa suits work on fun theory; and there's always the odd runner dressed as a giant chicken or a knight in armour running the London Marathon for charity, too. You do it instinctively in lots of ways already, and it doesn't have to be bells and whistles to be fun.

You don't even have to label it fun: you can call it tapping into the brain's pleasure zone instead. Because there's no getting away from our brains, and if something activates them in the area associated with pleasure and reward, then it's much more likely that they – and we – will want to participate in whatever type of pleasurable activity is being proposed.

Use reframing to release tension

Using reframing when things are not going that well can be extremely effective in both releasing tension and putting things in perspective. Note the *Pollyanna*-inspired[*] idea in the example below and think of all the times this could help us to react better when bad things happen. Anything, from dealing with accidents, children's homework issues or simply getting the whole family ready to go somewhere and out the door can all be potential stress points for which reframing works wonders.

> *Our family have a 'Glad' game. When something bad happens (say, paint spills all over the floor), we think about what might have been worse. E.g. 'Be glad that it was emulsion, so we could wipe it up more easily.'*

Go on, *chunk it*!

'Chunking' is a bit of a clunky word, there's no denying it, but it has a mechanical, purposeful feel.

> *The concept of 'chunking' has been very useful, even though it's a ghastly word! I can cheerfully report back that the 'c' word enabled me to clear up after a dinner for fourteen bit by bit, very happily.*

---

[*]   *Pollyanna* is a children's book by Eleanor Porter first published in 1913, which tells the story of Pollyanna, an orphan who tackles the difficulties of her life by attempting always to find something to be glad about, even when the situation looks very bleak.

By the time it was done, I discovered that it had really been quite pleasurable. In fact, I was almost sorry the job was over!

Brain-rewiring awareness: Framing and Chunking Strategies

Sometimes the simplest things can have the biggest impact: a straightforward piece of mental reframing that transforms a not-particularly enjoyable task like commuting into a more positive experience by claiming it as private time to read or listen to music; or thinking more carefully about the way in which you present ideas and how you can frame them to help to achieve your goals; or *re*framing an incoming message so that you are able to give a positive spin to what you first saw as a negative slight, by considering the context around the person who sent or said it; or just being open to the fact you might be misreading the delivery; or mentally chunking a daunting task into stages and, by doing this, making the impossible possible. It might just make that 500-page report more readable or the final ten laps of the pool a little more swimmable.

It's not an infallible method – some things are simply hard to deal with and reframing them doesn't make them go away – but it does offer a foothold of control over a bad situation, and the challenge of finding a good angle on a reframe can be helpful in itself. In a way, simply being able to stand back sufficiently to be able to ask yourself the question *'How can I reframe this unbearable situation to make it more bearable?'* or *'How can I reframe this person's negativity to keep me from losing my temper?'* or *'How can I chunk this task to make it easier?'* offers enough distance to diminish the tension. And it allows you to keep on turning the problem over in a search for a new way of solving it or living with it.

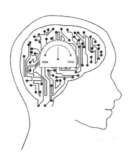

Summary of six brain nudges to encourage framing and chunking

1. Small changes in how information is presented or received can create a completely new perspective.
2. Frames can come in all shapes and sizes – some are simple, others can be wildly creative. The beauty of is you can choose the frame that works for you.
3. Use reframing to put yourself in someone's shoes and give them the benefit of the doubt.
4. Conceptualise framing/reframing as a simple piece of life-armoury you can carry with you and pull out any time you need it.
5. You can use reframing to build more positive (and kinder) reinterpretations of the past.
6. It is amazing what you can achieve if you take it step by step and break it down. 'Chunk it' is a great mantra.

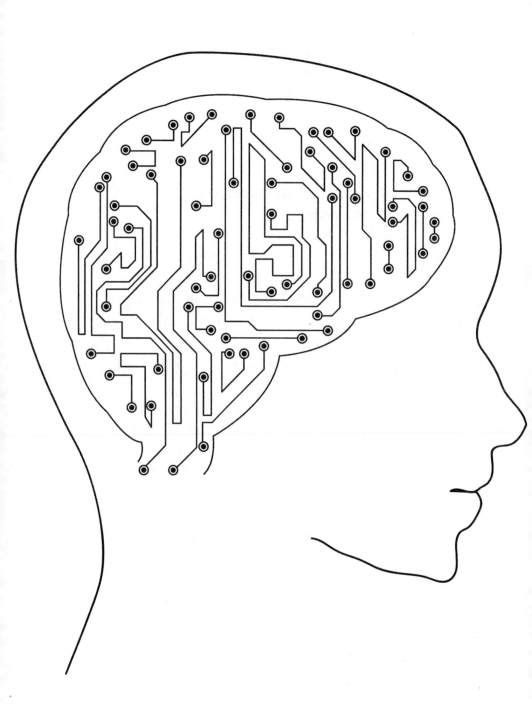

Section 4: Putting your newfound awareness into action and taking control

We have deep-dived into insights and explanations to show how our brains are wired with biases that make us behave, remember, see, say or believe things in ways we might not describe as wholly logical or useful. We have seen how we may not always act completely rationally, how we can become set in our ways of being and thinking, and how we might need a reboot to get free. While the biases play a critical part in our lives, making us cognitively efficient, stronger and potentially more protected, they also make us *less* conscious and *less* in control of our behaviour. The fact is, they don't always act in our best interest.

In the practical sections, we have demonstrated that just being aware of brain wirings and brain strategies, and knowing how they work, is an empowering experience that should make you better able to understand and interpret your own behaviour as well as the behaviour of others around you. During their participation in our project, the Brain Team reported that they were able to smooth the bumps out of their 24/7 lives, finding more joy and avoiding some of life's constant battles and tensions.

This knowledge is like a toolkit for your brain and arms you with some important insights and answers to issues such as:

⊙ Why it's so easy to operate on automatic mode and to be
   removed from the conscious moment, and how to wake up

⊙ How you can trigger a wave of positivity and stop a negative one

⊙ Why it feels so good to behave well to others and how to do it more often

⊙ Why it's easy for us all to get stuck in a rut and how you can climb out of it

⊙ Why we all have a tendency towards egocentricity and how important it is to manage this

⊙ Why we end up taking things for granted and how we can easily refresh our view of what's familiar

⊙ How the way we look at a problem or a goal can totally transform our ability to tackle it

You can use all this newfound awareness to make you a better coder and decoder of information. But if you are also able to begin to *act* more consciously on the awareness, to think about how to play to a particular piece of brain wiring or to try to block or counter another one, or about how to use a particular brain strategy to achieve a goal – however small – you will quickly see how small steps can lead to big changes. Here are six simple actions you can take right now:

1. By being more conscious of the now, and deliberately more mindful of times when you usually run on autopilot, you can make moments more memorable and longer lasting; you can slow down your life occasionally and find greater richness in the process.

2. By recognising how easy it is to start positive ripples, with the added excitement of not knowing where they might end, you can feel good about your days. By learning how to stop negative ripples, you can add calm. You can find something to feel good about every single day.

3. By acknowledging and understanding that our brains are wired with the tendency to resist change, and by consciously seeking to challenge this, you'll come to see that change isn't

always to be feared; it not only adds a little spice but can have a major positive impact on your life, encouraging a greater feeling of confidence and a sense of being in control.

4. By embracing the idea of egocentricity and opening your eyes to its influence on yourself, you'll see the world in a new light and have a heightened consciousness of your self-serving behaviours and those of others. You'll find huge benefits in fewer conflict situations, enhanced relationships and richer communications.

5. By confronting the adaptive process and trying to keep what's new feeling 'new' for longer – revitalising relationships and refreshing responses to possessions you've taken for granted – you'll be absorbed in a newfound appreciation of everything that's already present in your life and, with that, feel a deeper sense of contentment. You'll might even yearn for less and be happier with what you already have.

6. By becoming more sensitive and thoughtful about how you code or decode information, by taking time to approach problems or tasks in a more curious, evaluative manner and by adopting a steadier, step-by-step view of what is possible, you'll be inspired to conjure many positive and achievable strategies that will help to make your life a little less stressful.

So, keep thinking about the science, think about how the brain wirings might be impacting your world, and take your time with the thinking so as to fully engage with a process of awareness, absorption and action. Remember, even the smallest changes in thoughts and actions, and the smallest of rewiring nudges to your brain, can have an enormous and deeply satisfying impact on your life and on the lives of those around you.

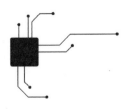

Appendix

Quick summary of the benefits of nudging your brain

Four Brain Nudges to Counter Autopilot

⊙ Become more conscious and aware of how autopilot operates across your life – the process alone is eye-opening.

⊙ Deepen your consciousness of a regular routine and see new things just by being more in the moment.

⊙ Engineer slight changes to your routines to jolt yourself into more active consciousness.

⊙ Take time out and do something that's not part of the plan.

Four Brain Nudges to Embrace the Reciprocity Bias

⊙ 'To give is to receive' – keep this mantra at the front of your mind.

⊙ Conceptualise the ripple effect spreading outwards. Who knows where it might end? Think about the idea of random acts of kindness – a chaos theory of kindness, a tidal wave of warmth. Start one today.

⊙ Use reciprocity to stop a negative wave.

⊙ The more you listen to someone and concentrate fully on what they are saying, the more likely they are to listen to and concentrate on you in return: a more rewarding experience for everyone involved.

Five Brain Nudges to Engage with Your Status Quo Bias

⊙ First gain awareness or increased consciousness of how it is operating and influencing your life – this new awareness alone will make you think differently.

⊙ Rocking the boat gently can have big impacts – just saying yes vs no, or picking up the phone and calling someone you've lost touch with, can rock your world.

⊙ Rocking the boat a little harder could well be life-changing, as one of our Brain Team discovered when increased spontaneity led to being asked on a date!

⊙ Try a status quo mantra for size – use 'Rock the boat', 'Seize the day', 'In boldness are magic and genius', 'Here's to the crazy ones', or even 'Shake the tree and knock the great sloth down on his ass.' Or you could just try 'Do one new thing today' and see if that works. Mantras are powerful, and they can help to empower us too.

⊙ Stop your boat from rocking! For some people who feel they already live their lives in turbulent waters, a little more calmness might be just the thing. You can say no to things.

Five Brain Nudges to Deflate Egocentric Bias

⊙ Be honest about this bias – anyone who feels they never or rarely display egocentric behaviour should, perhaps, ask themselves the question: is there not something just a little self-serving about considering yourself completely untainted by egocentricity?

⊙ The biggest insight for our Brain Team was how little people listen – and they tried hard to check this behaviour. A big takeaway is to listen harder and talk less, with longer silences. It can lead to deeper, more rewarding and less stressful conversations.

- ⊙ Come down off that pedestal – you're not so great.
- ⊙ Letting go of egocentricity can be restful. Releasing yourself from the need to compete can be a soothing and enjoyable experience.
- ⊙ You may need lots of little nudges to keep you focused on the goal of putting yourself second. The simple mantra 'It's not all about me' might help.

Six Brain Nudges to Combat the Normalising Bias

- ⊙ With conscious effort, you can derive new pleasure from what you've grown used to.
- ⊙ It's easy to let people become monochrome. Think about how you can restore colour to your relationships.
- ⊙ Reawaken old memories – think about the stories contained in the things around you and create a space for yourself to look, think and reminisce.
- ⊙ Awaken newfound appreciation – deprive yourself of something you take for granted.
- ⊙ Invest in experiences that have a rich and long-lasting value and that can set up a whole pattern of sustaining memories.
- ⊙ It will help if you make a list of the things that you take for granted but would like to refresh; the process itself will encourage you to take action.

Six Brain Nudges to Harness the Power of Framing and Chunking

⊙ Small changes in how information is presented or received can create a completely new perspective.

⊙ Frames can come in all shapes and sizes – some are simple, others can be wildly creative. The beauty is you can choose the frame that works for you.

⊙ You can also use reframing to put yourself in someone else's shoes and give them the benefit of the doubt.

⊙ Conceptualise framing/reframing as a simple piece of life-armoury you can carry with you and pull out any time you need it.

⊙ You can use reframing to build more positive (and kinder) reinterpretations of the past.

⊙ It is amazing what you can achieve if you take it step by step and break it down. 'Chunk it' is a great mantra.

List of Biases (A–Z)

Most of the biases discussed in this book operate outside conscious awareness, but once a bias has been identified, adjusting to accommodate its influence, reflecting on it or deliberately engaging or disengaging it is pretty straightforward. It's the *thinking* about biases we want you to focus on – to be generous, rather than miserly, in your consciousness of them and of their influence on your life.

Remember that simply heightening your awareness of these biases and, as a result, becoming more familiar and perhaps more comfortable with the idiosyncrasies of your brain, will have an impact on your everyday life, offering you the power to be more in control.

Anchoring Bias

Also known as the focusing effect, the anchoring bias describes our tendency in decision-making situations to look for some kind of reference point – we call it a mental 'anchor' – on which we can base a judgement or comparison. In a restaurant, we might look for a dish we know and, depending on that price, view the restaurant as expensive or good value. Sometimes we can rely too much on anchors when making a decision, but they offer us a choice structure and we can struggle to make a decision without one. Think about it: unless you have X to weigh up and to compare and contrast with, how will you ever be able to decide to buy Y?

Authority Bias

We grow up with authority figures who 'know best' – we start with our parents, then teachers. Later we listen to music selections by DJs we like, we trust doctors in white coats and are happy to believe that a play/book/movie is good because a newspaper critic rates it so. We are wired to seek out and believe authority figures; they make life easier, decision-making quicker. But it is worth questioning whether a person has a genuine authority or expertise or whether they're just voicing an opinion or even reading a script. We particularly need to watch out for 'authority seep' – when a person with genuine authority in one area is presented as an authority in another area in which they have absolutely no expertise.

Autopilot

Our brains are wired to make us cognitively efficient so that we don't waste time thinking about stuff we know how to do – we just get on and do it. What this means is that 95 per cent of our mental behaviour is subconscious and automatic – we simply aren't living as *consciously* as we think we are. In fact, scientists also estimate that around 45 per cent of our behaviour is habitual, i.e. we don't think about what we're doing, we just do it. Think about things you do today and each day of the week, and see what you can remember about them. It may come as a shock!

Availability Bias

This is a classic memory bias, which means we tend to estimate what is likely to happen according to what is more available in our memories – and the stuff we hold on to in our memories is biased towards things that are vivid, unusual or emotionally charged. It's why we might fear flying more than driving when statistically we are much more likely to be involved in a car accident than a plane crash.

Availability Cascade

Our brains are wired to think something is more believable if we hear it a lot (repeat something for long enough and it will come to seem true). This self-reinforcing cycle explains the development of certain kinds of collective beliefs. Think about how quickly gossip or speculation can begin to seem like fact.

Bias Blind Spot

This is a great starter for thinking about how our brains are wired, because the bias blind spot prevents us from being able to see our own biases and consequently supports our built-in tendency to fail to compensate for them. We are nonetheless likely to recognise evidence of biases in people around us, and it's no wonder disagreements occur and we still come out of an argument firmly believing ourselves to be right.

Chunking

When viewed as a whole, a task can seem daunting and unachievable. If we look at a task in its entirety we'll sometimes take the easy way out – *'It will take too long,'* we tell ourselves; *'It will be impossible to achieve in the time I have.'* Chunking allows us to break tasks down into pieces so we can tackle them one step at a time.

Commitment Bias

If you want to ensure that you stick to your plan, you have to commit. And the more tangible evidence you use to demonstrate your commitment, the more successful you're likely to be. Commitment bias can help to put paid to procrastination, inertia or impulsiveness. If you have a gym or jogging partner you don't want to let down – that's commitment bias; if you write notes to yourself and leave them on the fridge to try to stop yourself from snacking – that's commitment bias. It's about building a structure around a plan of action that will help you to keep to the plan.

Confirmation Bias

This is our tendency to search for, interpret and remember information in a way that confirms our existing beliefs or tells us something we want to hear. Not only that, but we are, unfortunately, more likely to accept facts uncritically if they *support* our theories or please us, and to insist on more evidence or better information when the facts seem not to be on our side.

Consistency Bias

Our brains are wired with the tendency to incorrectly remember our past attitudes and behaviour as resembling our *present* attitudes and behaviour, or to show a selective recall of past events to fit our current beliefs. So, we might forget we ever smoked or stayed out late at parties or handed our homework in late or did badly in an exam as teenagers – especially when we're talking to our own teenage children!

Consultation Paradox

Also called the interloper effect, this bias describes our tendency to regard third-party consultation as more objective and more valued than the opinions of those closer to us. 'You don't look fifty,' a consultant surgeon told a fifty-year-old woman at her first consultation, and it seemed so much more valid (and valuable) to her than if the same statement had been made by friends and family.

Denial Bias

This is the tendency to discount or disbelieve an important but uncomfortable fact. It is also a little like the ostrich effect, where we hide our heads in the sand to try to pretend something is not happening. It's linked to our inclination to accept facts which support our point of view and dismiss those that don't. The denial bias is all about self-protection – denial here is a defence mechanism.

Denomination Effect

Your brain is wired with a tendency to spend more money when it is denominated in small amounts than in large amounts. If we only have a £50 note, we'll be especially careful how – and indeed *if* – we spend it. It's much, much easier to spend a £5 note. It's a simple brain wiring that can help you spend less – but only if you make sure you only ever have high-value denomination notes on you.

Discounting the Future Bias

This is also known as time-inconsistent preferencing or, more simply, present bias, but what they all mean is that we are hardwired with a stronger preference to take proffered gains now and postpone the losses till later. We're also much more likely to value present gains over future potentially greater ones. It's easy to put off so many things till another day: the diet can wait; I'll join the gym next month; I'll start saving next year, etc. Basically, there's always tomorrow, and the You of tomorrow can handle the hard stuff while the You of today is very much into having your cake *and* eating it.

Distinction Bias

We have a tendency to view two options as more dissimilar when evaluating them together than when evaluating them separately. When seen together, small differences are much more in evidence than when viewed separately.

Dunning-Kruger Bias

We are all prey to the influence of the superiority bias, also known as the Dunning-Kruger bias. It makes us overestimate our skills and abilities while at the same time endowing us with a total lack of self-awareness that we're doing it.

Egocentric Bias

We are wired with a tendency to regard the rest of the world as if it were weighted to our own point of view, perspective or actions – as though we are the centre of our universe! Next time you are in a conversation, try to listen more carefully, and try not to spend 'listening' time thinking about what *you* are going to say next. Think about putting yourself second in other ways too: remind yourself you are not always right, apologise if you are in the wrong, take the blame, laugh at yourself, step off your pedestal.

Emotional Impact Bias

We are wired with a tendency to overestimate the duration or intensity of our future feelings. We think we will feel bad or happier for longer than we actually will. This bias demonstrates the power of the hot emotional moment to dominate our current and perceived future feelings.

False Consensus Effect

We are very likely to believe that what we think, and the opinions we hold, are shared by the majority. We don't think of ourselves as the odd ones out, nor do we want to be out of line with others. This may be because our friends and the people we spend time with are indeed like us, and we use the availability bias (the one that makes us depend on the things we can most easily recall when we are considering a specific concept) to deduce that many other people are similar to us too.

Focusing Illusion

This bias works hand-in-glove with the egocentric bias. In essence it blows the significance of whatever it is we're thinking about out of all proportion – we're thinking about it so it swells to have a sense of importance to the exclusion of all else. This can be useful to hang on to when you're anxious about something, whether it's something you've done or are about to do. Things usually diminish in significance if you leave them alone for a bit.

Framing Effect

The way in which information is presented to, or by, us (or received or interpreted by us) can completely change how it is viewed and the effect it creates. Providing the same information in a different way, or turning something on its head to see if there's another interpretation to be had, can change how we think about something and can make a tricky task more achievable. This is called framing and it's an incredibly powerful tool.

Fundamental Attribution Error

When we see someone doing something or behaving in a certain way, we assume it's because this is what they are like, i.e. because of their character, rather than considering that their actions have been prompted by external environmental or situational factors. So, if we see someone stub their toe or knock over a cup of coffee, we're likely to regard them as a little on the clumsy side, and if they're being a bit quiet at a social gathering, we might assume they are shy. Of course, if *we* stub our toe, it's not because we're clumsy, it's because someone else left the chair in a stupid place; and if we're quiet at a party, we're not shy, we're bored or worried about something.

Gamblers' Fallacy

This is a great example of irrational thinking, in that we believe that future probabilities are altered by past events, when in reality they remain unchanged by them. For instance, if the ball on a roulette wheel lands on red five times in a row, we might think that there is a higher chance it will land on black on the next spin of the wheel when there is no such probability. Similarly, after a long spell of dry weather, we might well begin to think we are 'due for some rain'.

Hawthorne Effect

This bias taps into social norms and refers to our tendency to perform differently when we know we are being observed by others. Because people are paying attention to us, we might behave either better or worse than if we felt ourselves to be unobserved. That nannyish phrase rings true: *'Don't pay them any attention and they'll soon stop doing it.'* And now we begin to understand the TV show *Big Brother*.

Herd Instinct

Our brains are wired with the tendency to adopt the opinions and follow the behaviour of the majority. The motivation can come from safety in numbers, pure laziness or to avoid conflict. This is also known as the social norm bias or the bandwagon effect.

Hindsight Bias

Past events appear to be more predictable than they actually were. Our brains like nothing better than order and patterns, so we have a tendency to filter our memories of past events through our present knowledge – like knowing how the novel turns out, because you've just read it.

Illusory Correlation Bias

This relates to the bias that makes us want to see order; so, we make patterns out of what is actually randomness and inaccurately link actions with effects. So, we might attribute meaning and purpose to something that could actually have happened just by chance. This also explains why we may often see shapes or even faces in things (clouds being a prime example). There's a word for this: pareidolia.

In-Group Bias

Our brains are wired with the tendency to give preferential treatment to those we perceive to be members of our own group. This has the benefit of making us feel better about ourselves and boosts our self-esteem. Think how you might have behaved in groups at school (sports teams, your class members, the choir) and how you behave with colleagues now. Notice how groups are formed and how non-'members' are regarded.

Irrational Escalation Bias

Also known as the sunk cost fallacy, this is the tendency for people to justify continued investment in a decision or action based on the previous investments (of money, time or effort) despite new evidence suggesting that the cost, starting today, of continuing the decision outweighs the expected benefit. You've been feeding a slot machine for twenty minutes and won nothing? Surely that means it's bound to pay out soon? You know the idioms – 'Throwing good money after bad', 'In for a penny, in for a pound'.

Just World Phenomenon

We are wired with a tendency to believe the world is 'just' and that people – others, particularly – 'get what they deserve'; that 'what goes around, comes around'. So, if we see something bad happening to someone else, we might find ourselves wondering what they have done to deserve it. It is easier to live in a world that we think is just, and where things happen for a purpose, than one where stuff just happens. Once again, it's our brains at work seeking to overlay some kind of logical explanation for things.

Licensing Effect

Licensing effect is when people allow themselves to do something bad after they've done something good. Or equally, metaphorically offset the 'carbon footprint' of their 'bad' behaviour by doing something 'good'. It's an effect that's described perfectly by Michael Rosenwald of the *Washington Post*: 'We drink Diet Coke – with Quarter Pounders and fries at McDonald's. We go to the gym – and ride the elevator to the second floor. We install tankless water heaters – then take longer showers. We drive SUVs to see Al Gore's speeches on global warming.'[1]

Look-Alike Bias

We can't help but be attracted to people who are like ourselves. We might, in jest, ask questions like 'Are they *people like us*?' to determine whether we will get on with strangers, but it's a question that's founded in reality. It could be an attraction based on physical or mental characteristics (and studies have shown that we are attracted to people who resemble us). By default, however, it can mean we reject people who are different from us. Birds of a feather, huh?

Loss Aversion

Also known as the endowment effect, this is our tendency to assign greater value and meaning to things we own, and to demand more recompense to give them up than we would be willing to pay for them. It's why free trials are so effective.

Need for Closure

We are wired with a tendency to prefer a recognisable end point or conclusion to almost anything. Our brains can feel a lack of satisfaction if something we have been involved with, or some issue we are dealing with, fades away without actually being concluded one way or another. This bias is particularly powerful in important matters. A lack of conclusion to something creates doubt, uncertainty and a sense of unfinished business.

Negativity Bias

We have a tendency for paying more attention to, and giving more weight to, negative experiences than positive ones. Bad things just seem to have a bigger impact on us and our brains. Who doesn't lean forward to listen better when a friend says, 'I'll tell you another terrible thing that happened…'? We have a tendency for negative comments to have a much weightier impact than positive ones. You know how you can let praise pass through you, but feel crushed by criticism? As American comedian Larry David says: 'A thousand positive remarks can slip by unnoticed, but one "You suck" can linger in your head for days.'

Normalising Bias

Our brains are wired to normalise things; it's a process that's part of adaptation and learning. What happens is that what was once fresh, new and exciting can quickly be translated into the norm by our normalising brains. Once something is 'learned' and adapted to, we no longer need to apply the level of conscious attention a new thing demands (from our brain's perspective, this would be akin to wasting unnecessary time *re*learning something we already know, so our brains tell us not to do this). We become accustomed to the people in our lives and to doing things or living with things, and we focus less and less on them so that they slip from the front-of-mind newness and associated high-emotional energy that was initially attached to them. Actively fight to keep the people and things you love fresh.

Optimism Bias

The human brain is wired with the tendency to be over-optimistic; it has been described as being 'hardwired for hope' and this may be a key wiring for progression and survival. Take the scene in *Monty Python's The Life of Brian* where they cheerily sing 'Always look on the bright side of life' as Brian is being crucified.

Overconfidence Bias

Here's another bias that derives from our positive evolutionary wiring and also relates to our self-belief and optimism biases: the tendency for excessive confidence that people can display in answers they give, such as in an exam or quiz. Psychological studies have shown that, in general, people rate their performance in exams as far better than they actually are (their ratings turn out to be wrong 40 per cent of the time). Whenever my son says that an exam has gone 'brilliantly', I immediately start to worry!

Peak-End Rule

We are wired with a tendency to remember an event based on two points: the emotional high-point – called 'the peak' – and 'the end'. In fact, much of the other detail can easily fade into the background. Now we know why party bags can often be the most memorable moment of the party for children. Never underestimate the importance of a *big* finish. Memories of experiences are strengthened by spiking elements within the experiences – good or bad. By contrast, if we have an overall good experience or a really long good experience with no specific high or low points, we probably won't remember it as well as we would if it had had one positive high-point somewhere along the way and then a big finish.

Planning Fallacy

This refers to our tendency to underestimate the time it will take to complete a task. This may be linked to the fact that we are hardwired to be optimistic and to be wishful thinkers, always looking for the silver lining in those clouds.

Priming

Priming is a simple and powerful way to influence behaviour. You can be primed to think or behave in a particular way (even though you aren't aware of it) by images, words, smells or the behaviour of others around you. You can be primed to be more polite, tidier, more determined to succeed or to take more risks. It's weird and powerful, and it happens all the time.

Reciprocity Bias

Our brains are wired so that we have the tendency to behave towards others as they behave to us. A positive action by you has a tendency to trigger a positive action in another person – so, start a positive wave! Likewise, a negative action has the power to create a negative wave. Think about implementing positive actions to counter negativity.

Restraint Bias

We are wired with a tendency to overestimate our ability to show restraint in the face of temptation. It is related to the cold–hot empathy gap. It is much easier to think about giving up alcohol or smoking when you are suffering from a mighty hangover on the morning after than it is in the middle of your happy, jolly, partying night before.

Retrenchment Bias

This is the tendency for deeply held beliefs to get stronger (and become more entrenched) when they are challenged by conflicting evidence. Also known as digging your heels in, perhaps.

Rosy Retrospection Bias

Our brains have a tendency to remember past events more positively than we actually felt them to be at the time. It's as if we have two experiencing selves – the actual experiencing self and the *remembering* self. After a certain amount of time has passed, we forget the annoying stuff – the cold, the rain, the long wait for the train, the irritating other people, etc. – and remember only the highlights, in all their glory. When we actually experienced the highlights, they would undoubtedly have been slightly diminished by the annoying stuff. *Ah, the good old days...* (My daughter looks back so fondly on her first festival experience: the drenching rain and mud viewed in retrospect add enormously to the adventure of the whole, when at the time she was just plain miserable. I know she was because she called me in the midst of it complaining about the hideousness of wet clothes, wet tent, wet sleeping bag, and so on.)

Scarcity Bias

We want what we can't have. It's as simple as that. If an online store lets you know that stock is low and there are only one or two remaining in this size or colour, you'll be nudged along to make that purchase. And if we are told there's a shortage of something because demand is particularly high for that thing, we'll want it more than if we're told the shortage is due to more practical considerations (late delivery, for instance). Amazon uses scarcity bias all the time – 'Only 2 units left'. Fashion labels use it regularly too. Some high-street retailers offer designer collaborations, releasing the garments at a precise time online and in selected stores, and in limited runs, all driven by a tantalising preview campaign and a word-of-mouth rising anxiety about being beaten to it online or on the phone. As consumers, we are driven partly by 'anticipated regret' as we imagine how fed up we'll be when stock runs out or we miss our opportunity.

Self-fulfilling Prophecy

We have a tendency to behave in ways that will elicit the results we are looking for or anticipating. So, a person who says 'I'm probably going to have a lousy day' might subconsciously alter their behaviour so that such a prediction is fulfilled. And by the same token, a person who addresses stuff in a positive way by saying 'I'm going to have a great day' might act in ways that will actually make this prediction true. It's attitude and *then* behaviour, in this case.

Self-serving Bias

We have an in-built tendency to claim more responsibility for our successes than for our failures and we judge ourselves more favourably than others on most dimensions. Research shows that just about everyone thinks they are more competent than their co-workers, friendlier than the general public, more intelligent than their peers, less prejudiced than their neighbours, younger-looking than other people the same age, better drivers than most people they know. We believe if we do well, it's all down to us – our skills and our tenacity – and if we do badly, someone or something else is to blame. Our brains are wired like this to make us feel good about ourselves and to protect our self-esteem.

Status Quo Bias

We are predisposed to want to keep things the same; to maintain the status quo and to avoid change. This means we have a tendency to defend and even boost the status quo, whatever the context. We're oriented not to rock the boat and rarely stick our heads over the parapet. Think a little about all the everyday routines you have, and maybe shake one of them up a little – who knows what might happen!

Texas Sharpshooter Fallacy

This is also called the clustering illusion and refers to the concept of firing random shots at a barn door, then drawing a circle around the best group of holes and declaring this to be the target. It's a prime example of moving the goalposts.

Notes

Section 1

1. Janet A. Schwartz and Dan Ariely, 'Life is a Battlefield', *Independent Review*, Vol. 20, 3 (Winter 2016), pp. 377–382.

2. John A. Bargh, 'The Automaticity of Everyday Life', in Robert S. Wyer (Ed.), *The Automaticity of Everyday Life: Advances in Social Cognition, Vol. X*, Lawrence Erlbaum Associates, New Jersey, 1997.

3. Sheena Iyengar, *The Art of Choosing*, Little, Brown, London, 2010.

4. Bas Verplanken and Wendy Wood, 'Interventions to Break and Create Consumer Habits', *Journal of Public Policy & Marketing*, Vol. 25, 1 (2006), pp. 90–103, cited in Charles Duhigg, *The Power of Habit: Why We Do What We Do in Life and Business*, Random House, London, 2012.

5. Bargh, 'The Automaticity of Everyday Life'.

6. John Dewey, *How We Think*, D. C. Heath & Co., Boston, 1910.

7. Amanda Staveley, 'What She Said: Amanda Staveley Answers Your Workplace Dilemma', *Sunday Times Style* magazine, 27 May 2018.

8. Richard H. Thaler and Cass R. Sunstein, *Nudge: Improving Decisions About Health, Wealth, and Happiness*, Yale University Press, New Haven, 2008.

9. Daniel Kahneman, *Thinking, Fast and Slow*, Farrar, Strauss and Giroux, New York, 2011.

10. Max Ernest-Jones, Daniel Nettle and Melissa Bateson, 'Effects of Eye Images on Everyday Cooperative Behavior: A Field Experiment', *Evolution and Human Behavior*, Vol. 32 (2011), pp. 172–8.

11. These are the words of the author's daughter.

12. https://www.realmenrealstyle.com/looks-matter-men/

13. John A. Bargh, Mark Chen and Lara Burrows, 'Automaticity of Social Behavior: Direct Effects of Trait Construct and Stereotype Activation on Action', *Journal of Personality and Social Psychology*, Vol. 71, 2 (1996), pp. 230–44.

14. Rachel Herz, *Why You Eat What You Eat: The Science Behind Our Relationship With Food*, W. W. Norton, New York, 2018.

15. Lawrence E. Williams and John A. Bargh, 'Experiencing Physical Warmth Promotes Interpersonal Warmth', *Science*, Vol. 322, 5901 (2008), pp. 606–7.

16. Ibid.

17 Bargh, Chen and Burrows, 'Automaticity of Social Behavior'.

18 http://washington.providence.org/senior-care/mount-st-vincent/services/child-care/

19 *Old People's Home for 4 Year Olds*, TV, Channel 4, August 2017.

20 Sara L. Bengtsson, Raymond J. Dolan and Richard E. Passingham, 'Priming for Self-Esteem Influences the Monitoring of One's Own Performance', *Social Cognitive and Affective Neuroscience*, Vol. 6, 4 (2011), pp. 417–25.

21 Brett W. Pelham, Matthew C. Mirenberg and John T. Jones, 'Why Susie Sells Sea-Shells By the Sea Shore: Implicit Egotism and Major Life Decisions', *Journal of Personality and Social Psychology*, Vol. 82, 4 (2002), pp. 469–87.

22 The Doctors' Names List: http/www.u.arizona.edu/~stoddard/doctor.htm

23 http://www.richardwiseman.com/quirkology/new/USA/Experiment_names.shtml

24 David N. Figlio, 'Names, Expectations and the Black-White Test Score Gap', National Bureau of Economic Research, 2005.

25 Rob W. Holland, Merel Hendriks and Henk Aarts, 'Smells Like Clean Spirit: Nonconscious Effects of Scent on Cognition and Behaviour', *Psychological Science*, Vol. 16, 9 (2005), pp. 689–93.

26 Noel E. Brick, Megan J. McElhinney and Richard S. Metcalfe, 'The Effects of Facial Expression and Relaxation Cues on Movement Economy, Physiological, and Perceptual Responses During Running', *Psychology of Sport and Exercise*, Vol. 34 (2018), pp. 20–28.

27 John A. Bargh and Tanya L. Chartrand, 'The Unbearable Automaticity of Being', *American Psychologist*, Vol. 54, 7 (July 1999).

28 Ibid.

29 Birte Englich, Thomas Mussweiler and Fritz Strack, 'Playing Dice With Criminal Sentences: The Influence of Irrelevant Anchors on Experts' Judicial Decision Making', *Personality and Social Psychology Bulletin*, Vol. 32, 2 (2006), pp. 188–200.

30 Scott Eidelman, Chris S. Crandall, Jeffrey Goodman et al., 'Low-Effort Thought Promotes Political Conservatism', *Personality and Social Psychology Bulletin*, Vol. 38, 6 (2012), pp. 808–20.

31 'The World Urgently Needs Critical Thinking, Not Gut Feeling', *New Scientist*, 13 December 2017.

32 John Bargh, *Before You Know It: The Unconscious Reasons We Do What We Do*, Touchstone, New York, 2017.

33 Dan Ariely and George Loewenstein, 'The Heat of the Moment: The Effect of Sexual Arousal on Sexual Decision Making', *Journal of Behavioral Decision Making*, Vol. 19, 2 (2006), pp. 87–98.

34 Thomas Schelling, 'The Intimate Contest for Self-Command', *Choice and Consequence: Perspectives of an Errant Economist*, Harvard University Press, Cambridge, Massachusetts, 1984.

35 George Loewenstein, 'Hot-Cold Empathy Gaps and Medical Decision Making', *Health Psychology*, Vol. 24, 4S (2005), pp. S49–56.

36 M. L. Slevin, L. Stubbs, H. J. Plant et al., 'Attitudes to Chemotherapy: Comparing Views of Patients with Cancer with Those of Doctors, Nurses, and General Public', *British Medical Journal*, Vol. 300 (1990), pp. 1458–60.

37 Itamar Simonson and Amos Tversky, 'Choice in Context: Tradeoff Contrast and Extremeness Aversion', *Journal of Marketing Research*, Vol. 29, 3 (1992), pp. 281–95.

38 David A. Schkade and Daniel Kahneman, 'Does Living in California Make People Happy? A Focusing Illusion in Judgments of Life Satisfaction', *Psychological Science*, Vol. 9, 5 (1998), pp. 340–6.

39 https://worldpopulationreview.com/country-rankings/murder-rate-by-country

40 J. Edward Russo and P. J. H. Schoemaker, *Winning Decisions: Getting It Right the First Time*, Crown Business, New York, 2001.

41 In *Silent Night*, Series 15, Episode 5, BBC Radio 4.

42 Amos Tversky and Daniel Kahneman, 'The Framing of Decisions and the Psychology of Choice', *Science*, Vol. 211, 4481 (1981), pp. 453–8.

43 J. B. Detweiler, B. T. Bedell, P. Salovey et al., 'Message framing and sunscreen use: gain-framed messages motivate beach-goers', *Health Psychology*, Vol. 18, 2 (1999), pp. 189–96.

44 Ernst Fehr and Simon Gächter, 'Fairness and Retaliation: The Economics of Reciprocity', *Journal of Economic Perspectives*, Vol. 14, 3 (2000), pp. 159–181.

45 Summer Allen and Jill Suttie, 'How Our Brains Make Us Generous', *Greater Good Magazine*, 21 December 2015.

46 Jamil Zaki and Jason P. Mitchell, 'Equitable decision making is associated with neural markers of intrinsic value', *PNAS*, Vol. 108, 49 (2011), pp. 19761–6.

47 Danusha Laméris, 'Small Kindnesses', in James Crews (Ed.), *Healing the Divide: Poems of Kindness and Connection*, Green Writers Press, 2019.

48 Supermodel Kate Moss in an interview in *WWD* (13 November 2009) in answer to the question, 'Do you have any lifestyle mottos?'

49 Loretta Graziano Breuning, 'Your Neurochemical Self: When Someone Pushes Your Buttons, Know Your Own Buttons', Psychologytoday.com, 9 July 2011.

50 H. Cai, Y. Chen and H. Fang, 'Observational Learning: Evidence from a randomized natural field experiment', *American Economic Review*, Vol. 99, 3 (2009), pp. 864–82.

51 S. E. Asch, 'Effects of group pressure upon the modification and distortion of judgments', in H. Guetzkow (Ed.), *Groups, Leadership and Men*, Carnegie Press, Pittsburgh, 1951, pp. 177–90.

52 Gregory S. Berns, Jonathan Chappelow et al., 'Neurological Correlates of Social Conformity and Independence During Mental Rotation', *Biological Psychiatry*, Vol. 58, 3 (2005), pp. 245–53.

53 Vasily Klucharev, Kaisa Hytönen, Mark Rijpkema et al., 'Reinforcement Learning Signal Predicts Social Conformity', *Neuron*, Vol. 61, 115 (2009), pp. 140–51.

54 https://littlethings.com/lifestyle/social-conformity-experiment

55 https://www.facebook.com/Prudential/videos/945170852264591/

56 Bargh and Chartrand, 'The Unbearable Automaticity of Being'.

57 David Leonhardt, 'Two Big Misperceptions', *New York Times*, 17 January 2018.

58 Jan B. Engelmann, C. Monica Capra, Charles Noussair et al., 'Expert Financial Advice Neurobiologically "Offloads" Financial Decision-Making Under Risk', *PLoS ONE*, Vol. 4, 3 (2009), pp. 1–14.

59 https://www.gov.uk/government/publications/e-cigarettes-an-evidence-update, August 2015.

60 Collaborative Group on Hormonal Factors in Breast Cancer, 'Type and timing of menopausal hormone therapy and breast cancer risk', *Lancet*, Vol. 394, 10204 (28 September 2019).

61 Tom Whipple, 'Benefits of HRT "still outweigh risks"', *The Times*, 31 August 2019.

62 Robert Webb, 'Peep Show Star Robert Webb on Grim Flatshares, Domestic Duties and the Joy of Napping', *Sunday Times*, 3 June 2018.

63 David G. Myers, *The Inflated Self: Human Illusions and the Biblical Call to Hope*, Seabury Press, New York, 1980.

64 David Myers, 'Humility: Theology Meets Psychology', *Reformed Review*, Vol. 48 (1995), pp. 195–206.

65 Justin Kruger and David Dunning, 'Unskilled and unaware of it: how difficulties in recognizing one's own incompetence lead to inflated self-assessments', *Journal of Personality and Social Psychology*, Vol. 77, 6 (1999), pp. 1121–34.

66 Joyce Ehrlinger and David Dunning, 'How chronic self-views influence (and potentially mislead) estimates of performance', *Journal of Personality and Social Psychology*, Vol. 84, 1 (2003), pp. 5–17.

67 https://www.bbc.co.uk/news/health-43077465

68 Hillary Rodham Clinton, *What Happened*, Simon & Schuster, New York, 2017.

69 Katty Kay and Claire Shipman, *The Confidence Code: The Science and Art of Self-Assurance – What Women Should Know*, Harper Business, New York, 2014.

70 @janegarvey1, Jane Garvey, Twitter, 29 May 2019.

71 Tali Sharot, *The Optimism Bias: A Tour of the Irrationally Positive Brain*, Pantheon, New York, 2011.

72 *The Green Book: Central Government Guidance on Appraisal and Evaluation (2020)*, HM Treasury, London, 2020, available at gov.uk.

73 Sam Coates and Lucy Fisher, 'Boris Johnson and Theresa May Clash Over the Blame for Donald Trump's Cancelled Visit', *The Times*, 13 January 2018.

74 Oliver Burkeman, 'This Column Will Change Your Life: Hindsight – It's Not Just for Past Events', *Guardian*, 10 May 2014.

75   'Strategic Decisions: When Can You Trust Your Gut?', *McKinsey Quarterly*, McKinsey & Company, March 2010.

76   J. D. Trout, 'The Psychology of Discounting: A Policy of Balancing Biases', *Public Affairs Quarterly*, Vol. 21, 2 (2007), pp. 201–220.

77   https://news.uchicago.edu/story/dissertation-write-provides-camaraderie-and-focus-graduate-students

78   M. S. Pallak, D. A. Cook and J. J. Sullivan, 'Commitment and Energy Conservation', *Applied Social Psychology Annual*, Vol. 1 (1980), pp. 235–54.

79   Barry M. Staw, 'Knee-deep in the Big Muddy: A Study of Escalating Commitment to a Chosen Course of Action', *Organizational Behavior and Human Performance*, Vol. 16, 1 (1976), pp. 27–44.

80   http://sidewiseinsights.blogspot.com/2010/03/

81   R. F. Baumeister, E. Bratslavsky et al., 'Bad Is Stronger Than Good', *Review of General Psychology*, Vol. 5, 4 (2011), pp. 323–370.

82   gretchenrubin.com/2007/01/hedonic_adaptat/

83   Stefanie Brassen, Matthias Gamer and Christian Büchel, 'Anterior Cingulate Activation is Related to a Positivity Bias and Emotional Stability in Successful Aging', *Biological Psychiatry*, Vol. 70, 2 (2011), pp. 131–7.

84   Karl Popper, *Conjectures and Refutations: The Growth of Scientific Knowledge*, Routledge and Kegan Paul, London, 1963.

85   Karl Popper, 'Reply to my Critics' in P. A. Schilpp (Ed.), *The Philosophy of Karl Popper*, Open Court, La Salle, Illinois, 1974.

86   David Leonhardt, *New York Times*, 1 August 2017.

87   J. M. Darley and P. H. Gross, 'A hypothesis-confirming bias in labeling effects', *Journal of Personality and Social Psychology*, Vol. 44, 1 (1983), pp. 20–33.

88   Jonathan Leake, 'Science Has Dim View of Brexit Voters' Brains', *Sunday Times*, 10 March 2019.

89   Thaler and Sunstein, *Nudge*.

90   Terrance Odean, 'Are Investors Reluctant to Realize their Losses?', *Journal of Finance*, Vol. 53, 5 (1998), pp. 1775–98.

91   Devin G. Pope and Maurice E. Schweitzer, 'Is Tiger Woods Loss Averse? Persistent Bias in the Face of Experience, Competition, and High Stakes', *American Economic Review*, Vol. 101, 1 (2011), pp. 129–57.

92   Ibid.

93   Daniel Kahneman, Jack L. Knetsch and Richard H. Thaler, 'Experimental Tests of the Endowment Effect and the Coase Theorem', *Journal of Political Economy*, Vol. 98, 6 (1990), pp. 1325–48.

94   Dan Ariely, *Predictably Irrational: The Hidden Forces That Shape Our Decisions*, HarperCollins, New York, 2008.

95   Figures from the now-defunct Department of Trade and Industry's annual *Home Accident Surveillance System 2001*.

96   https://davidmyers.org/articles/do-we-fear-the-right-things

97   David McRaney, *You Are Not So Smart*, Gotham Books, New York, 2011.

98 Clinton, *What Happened.*

99 Daniel Kahneman, Barbara L. Fredrickson and Charles A. Schreiber, 'When More Pain Is Preferred to Less: Adding a Better End', *Psychological Science,* Vol. 4, 6 (1993), pp. 401–5.

100 Kahneman, *Thinking, Fast and Slow.*

101 Emily Pronin and Matthew B. Kugler, 'Valuing Thoughts, Ignoring Behaviour: The Introspection Illusion as a Source of the Bias Blind Spot', *Journal of Experimental Social Psychology*, Vol. 43, 4 (2007), pp. 565–78.

102 Ibid.

103 Emily Pronin, Daniel Y. Lin and Lee Ross, 'The Bias Blind Spot: Perceptions of Bias in Self Versus Others', *Personality and Social Psychology Bulletin*, Vol. 28, 3 (2002), pp. 369–81.

104 Ibid.

Section 2

1   Reported in *The Times*, 10 March 2018.

2   Noah. J. Goldstein, Robert B. Cialdini and Vladas Griskevicius, 'A Room with a Viewpoint: Using Social Norms to Motivate Environmental Conservation in Hotels', *Journal of Consumer Research*, Vol. 35, 3 (2008), pp. 472-82.

3   K. Baca-Motes, A. Brown, A. Gneezy et al., 'Commitment and behavior change: Evidence from the field' *Journal of Consumer Research*, Vol. 39, 5 (2013), pp. 1070–84.

4   '2008 Corporate Responsibility Report', The Walt Disney Company, 2008.

5   Correct on 31 October 2021.

6   http://hospicefoundation.ie/wp-content/uploads/2020/12/Design-Dignity-Guidelines-Irish-Hospice-Foundation-2020.pdf

7   Crawford Hollingworth, inventor of Brainy Bike Lights.

8   https://www.bi.team/about-us/who-we-are/

9   https://www.bi.team/wp-content/uploads/2017/10/BIT_Update-16-17_E_.pdf

10 OECD, *Behavioural Insights and Public Policy: Lessons from Around the World*, OECD Publishing, Paris

11 https://www.kingsfund.org.uk/blog/2015/07/how-much-has-generic-prescribing-and-dispensing-saved-nhs

12 https://accessiblemeds.org/resources/blog/2017-generic-drug-access-and-savings-us-report

13 http://nudgeunit.upenn.edu/projects/using-default-options-increase-generic-medication-prescribing-rates

14 https://ldi.upenn.edu/news/nejm-spotlights-penns-history-making-nudge-unit

15 Robert B. Cialdini, *Influence: The Psychology of Persuasion*, W. Morrow and Company, New York, 1993.

16 Department for Environment, Food & Rural Affairs and the Right Hon Rory Stewart OBE, 'Plastic Bag Numbers Rise for the Fifth Year', www.gov.uk, 24 July 2015.

17 https://www.gov.uk/government/publications/carrier-bag-charge-summary-of-data-in-england/single-use-plastic-carrier-bags-charge-data-in-england-for-2018-to-2019

18 L. L. Shu, N. Mazar, F. Gino et al, 'Signing at the beginning makes ethics salient and decreases dishonest self-reports in comparison to signing at the end', *Proceedings of the National Academy of Science,* Vol. 109, 38 (2012), pp. 15197–15200.

19 Peter John, Sarah Cotterill, Alice Moseley et al., *Nudge, Nudge, Think, Think: Experimenting with Ways to Change Citizen Behaviour*, Bloomsbury Academic, London, 2013.

20 https://casaa.org/wp-content/uploads/Behaviour-Change-Insight-Team-Annual-Update_acc.pdf

21 Michael Cooper, 'From Obama, the Tax Cut Nobody Heard Of', *New York Times*, 18 October 2010.

22 http://foreignpolicy.com/2018/04/03/life-inside-chinas-social-credit-laboratory/

Section 3

1 Joshua Foer, *Moonwalking with Einstein: The Art and Science of Remembering Everything*, Penguin, New York, 2012.

2 https://www.alz.co.uk/research/WorldAlzheimerReport2014.pdf

3 James K. Rilling, Gregory S. Berns et al., 'A Neural Basis for Social Cooperation', *Neuron*, Vol. 35, 2 (July 2002), pp. 395–405.

4 Fehr and Gächter, 'Fairness and Retaliation'.

5 http://art.tfl.gov.uk//media/michael-landy-acts-of-kindness

6 Cialdini, *Influence*.

7 Ray Bradbury, *Fahrenheit 451*, Ballantine Books, New York, 1953.

8 https://www.youtube.com/watch?v=YBJAvi3A0H8

9 William Samuelson and Richard Zeckhauser, 'Status Quo in Decision Making', *Journal of Risk and Uncertainty*, Vol. 1, 1 (1988), pp. 7–59.

10 Gaurav Suri, Gal Sheppes, Carey Schwartz et al., 'Patient Inertia and the Status Quo Bias: When an Inferior Option Is Preferred', *Psychological Science,* Vol. 24, 9 (2013), pp. 1763–69.

11 Stephen M. Fleming, Charlotte L. Thomas and Raymond J. Dolan, 'Overcoming status quo bias in the human brain', *PNAS*, Vol. 107, 13 (2010), pp. 6005–9.

12 Suri et al., 'Patient Inertia and the Status Quo Bias'.

13 Matthew Syed, 'Failure can be the route to success for today's stressed schoolchildren', *The Times*, 30 June 2018.

14 Ibid.

15 https://www.ted.com/talks/sir_ken_robinson_do_schools_kill_creativity

16 https://onthespike.com/

17 Michael Ross and Fiore Sicoly, 'Egocentric Biases in Availability and Attribution', *Journal of Personality and Social Psychology*, Vol. 37, 3 (1979), pp. 322–36.

18 David G. Myers, *The Inflated Self: Human Illusions and the Biblical Call to Hope*, Seabury Press, New York, 1980

19 Ibid.

20 @MarianKeyes, Marian Keyes' World, Twitter, 9 July 2019.

21 Philip Brickman and Donald T. Campbell, 'Hedonic Relativism and Planning the Good Society', in M. H. Appley (Ed.), *Adaptation Level Theory: A Symposium*, Academic Press, New York, 1971, pp. 287–302.

22 David G. Myers, *The American Paradox: Spiritual Hunger in an Age of Plenty*, Yale University Press, New Haven, 2000.

23 Robert E. Lane, *The Loss of Happiness in Market Democracies*, Yale University Press, 2000.

24 Giles Coren, 'If you think reality TV's bad, try the Game of Life', *The Times*, 29 June 2019.

25 Robert Biswas-Diener, Ed Diener and Maya Tamir, 'The Psychology of Subjective Well-Being', *Daedalus*, Vol. 133, 2 (2004), pp. 18–25.

26 Dan Ariely, *The Upside of Irrationality*, Harper, New York, 2010.

27 Claudia Dreifus, 'The Smiling Professor', *New York Times*, 22 April 2008.

28 Thomas Gilovich, Amit Kumar and Lily Jampol, 'A Wonderful Life: Experiential Consumption and the Pursuit of Happiness', *Journal of Consumer Psychology*, Vol. 25, 1 (2015), pp. 152–65.

Appendix

1 Michael Rosenwald, 'Why going green won't make you better or save you money', *Washington Post*, 18 July 2010.

Index

Unbound is the world's first crowdfunding publisher, established in 2011.

We believe that wonderful things can happen when you clear a path for people who share a passion. That's why we've built a platform that brings together readers and authors to crowdfund books they believe in – and give fresh ideas that don't fit the traditional mould the chance they deserve.

This book is in your hands because readers made it possible. Everyone who pledged their support is listed below. Join them by visiting unbound.com and supporting a book today.

Mark Abell
Paul Allen
Nicola Alloway
Marta Alves Simões
Peter Anderson
Grant Cameron Anthony
Nesher Asner
Tim Atkinson
Vikki Baker
Darren Ball
Derren Ball
Ginny Battcock
Humphrey Battcock
Juli and Simon Beattie
Katy Beatton
David Beeson
Heather Binsch
Noreen Blanluet
Susan Booth
Sophie Bowers
Nicky Brown
Charlie Buckley
Rachel Burden
Emily Burgardt
Baz Butcher

Mike Butcher
Ian Calcutt
Leo Campbell
William Campion
Mark Cannell
Bobi Carley
Oliver Cary
Joanne Chittenden
Sally Collins
Richard Connor
Patrick Corr
Marie-Charlotte Courouble
Dan Cresta
Miranda Creswell
Sarah Crisp
Julia Croyden
David Cunningham
Melanie Cunningham
Sheila Cunningham
Michael Daniels
Polly Davidson
Dirk DeBoi
Justin Deering
Rosie Denlegh-Maxwell
Patric ffrench Devitt

Julie Doleman
Michail Dim. Drakomathioulakis
Catherine Driscoll
Mark Earls
Tim Edwards
C.J Esplin
Alison Falconer
Helen Farmer
Dr A.L.M. Fingret
Andrew Fingret and
   Kate Burdette
Claire Firth
Nicholas Foster
Clare Fowler
Julie Gibbon
Mark Given
Fergus Gleeson
Lisa Goodchild
George Goodfellow
Dianna Goodwin
Claire Grant
Gemma Greaves
Nicole Greenfield-Smith
David Gye
Maximilian Hardie
Melissa Harding
Zoe Harkness
Maddie Hollingworth
Emma Harris
Ralph Helm
Joey Helsby
Amanda Hollingworth
Harry Hollingworth
Annabel Holroyd
Alex Howard
George Howard
Jane Howard
Joshua Howlett
Catherine Huck
Pip Jamieson
Toby Jeffries
Wendy Jeskins
Benjamin Jones
Karen Jones

Sarah Jones
K & KJ Associates GmbH
Dan Kieran
Kat Kline
Claire Koryczan
Caroline Lamb
Susie Lambert
Martin Lane
Charlie Lee-Potter
Emilia Leese
David Leonard
Tamara Littleton
Mari Lloyd
Sarah Lloyd
Jonathan Lloyd-Jones
Alice Lyon
Rob MacAndrew
Siobhan Mackenzie
Tracey MacLeod
Stephen Maher
Catt** Makin
Ashvin Malhotra
Glenn Manoff
Anthony Maplesden
Joshua Martin
Mike Maurer
Dawood Mayet
Lucy McCahon
Christene McCauley
Nicola McLaren
Tomeu Miralles
Sonia Misak
John Mitchinson
Linda Monckton
Richard Montagu
Carol Montgomery
Carlo Navato
Evan Nesterak
John New
Al Nicholson
Lara Nicoll
Lasse Nielsen
Niels Aagaard Nielsen
Alexander Nirenberg

Terry O'Brien
Jackie O'Sullivan
Nic Parsons
Stephen Paton
Justin Pollard
Laura Pollard
Elaine Pretty
Maddy Price
Oliver Rees
Brenda Reilly
Guy Richardson
Sophie Ridley
Lucille Rigden
Kinga Rona-Gabnai
Giles Rooney
Phillipa Rooney
Bernard Ross
Shafik Saba
Christoph Sander
Lyni Sargent
Carol Sayles
Tricia Schmid
David Scott
Bill Scott-Kerr
Sally Smallman
Toni Smerdon
Sarah Smith
Wendy Staden
Mark Stahlmann
Paul Stather-Hooper

Philip Stinson
Callum Strachan
Christopher Stuart
Rory Sutherland
Kate Swinburn
Maisie Taylor
Kelly Teasdale
Paula Tebay
Emma Thimbleby
Jens Tholstrup
Mark Thompson and
   Jane Blumberg
Frank Thomson
Anne Tomlinson
Gerry Tomlinson
Maggie Tomlinson
Mandy Tomlinson
Allison Turner
Cat Turner
Eleanor Tweddell
Sophia Ufton
Sonja van Amelsfort
Emmie Van Biervliet
Mark Vent
Sir Harold Walker
Chee Lup Wan
John Wates
Alexandra Welsby
Bridget Wood
Stephen Usins Yeardley